FROM MOLTKE TO BIN LADEN

THE RELEVANCE OF DOCTRINE IN THE CONTEMPORARY MILITARY ENVIRONMENT

Albert Palazzo

Land Warfare Studies Centre
Canberra
September 2008

National Library of Australia Cataloguing-In-Publication Entry

Palazzo, Albert, 1957-

From Moltke to Bin Laden : the relevance of doctrine in the contemporary military environment

Bibliography.

ISBN 9780642296849 (pbk.)

Study paper (Land Warfare Studies Centre (Australia)) ; no. 315.

Subjects: Military doctrine.
Psychology, Military.
War (Philosophy)
War--Moral and ethical aspects.

172.42

Land Warfare Studies Centre Study Papers

ISSN 1442-8547

Study papers produced by the Land Warfare Studies Centre are vehicles for progressing professional discussion and debate concerning military strategy, particularly the application of land warfare concepts and capabilities to the security of Australia and its interests. Study papers are intended to provide comprehensive treatment of their subject matter at the time of publication.

Series Editor: Michelle Lovi

Land Warfare Studies Centre

The Australian Army established the LWSC in July 1997 through the amalgamation of several existing staffs and research elements.

The charter of the LWSC is to promote the wider understanding and appreciation of land warfare; provide an institutional focus for applied research into the use of land power by the Australian Army; and raise the level of professional and intellectual debate within the Army. The LWSC fulfils these roles through a range of internal reports and external publications; a program of conferences, seminars and debates; and contributions to a variety of professional, academic and community forums. Additional information on the centre may be found on the Internet at <http://www.defence.gov.au/army/lwsc/>.

Comment on this paper is welcome and should be forwarded in writing to:

The Director, Land Warfare Studies Centre
Ian Campell Road, Duntroon ACT 2600
AUSTRALIA

Telephone: (02) 6265 9890
Facsimile: (02) 6265 9888
Email: <lwsc.publications@defence.gov.au>

About the Author

Dr Albert Palazzo is a Senior Research Fellow at the Australian Army's Land Warfare Studies Centre in Canberra. His PhD is from The Ohio State University and his BA and MA are from New York University. He has written widely on warfare in the modern age and on the Australian Army in particular. His many publications include: *Seeking Victory on The Western Front: The British Army & Chemical Warfare in World War I*; *The Australian Army: A History of its Organisation, 1901-2001*; *Defenders of Australia: The Third Australian Division*; *Battle of Crete*; *The Royal Australian Corps of Transport*; and *Australian Military Operations in Vietnam*.

Acknowledgments

The author wishes to thank the Trustees of the Liddell Hart Centre for Military Archives for permission to quote from the holdings of the Liddell Hart Centre for Military Archives. The author also wishes to acknowledge the support of the Australian Research Council in the research of this monograph.

I have come to believe that doctrine
has often tended to inhibit the learning process.[1]

---

1 Jay Luvaas, *The Military Legacy of the Civil War: The European Inheritance*, University Press of Kansas, Lawrence, Kansas, 1988, p. xvii.

Contents

Introduction

Contemporary military organisations regard the possession, dissemination and inculcation of a doctrine of warfighting as an absolute necessity. This doctrine plays a central role in the development of warfighting techniques and is an important—if not the most important—factor in determining a force's capabilities and effectiveness. Doctrine *per se* serves as a symbol of professionalism and is a reliable tool through which to mandate uniformity of purpose. Military practitioners can also use doctrine to interpret the art of war, identify necessary changes and implement desired adaptations.

The lessons of the past reinforce the vital role of doctrine in the articulation of modern military power. Well conceived doctrine acts as a force multiplier and can provide a combatant with an advantage over an opponent—even one that is similarly armed and organised or which possesses greater mass and resources. Conversely, the historical record is replete with examples of defeated nations whose armed forces implemented inferior or flawed doctrines, and whose leaders, soldiers and citizenry ultimately bore the consequences.

Yet, in their allegiance to doctrine, today's military professionals and defence thinkers are prone to overlook its limitations. They tend not to consider its relatively recent origins nor that it is a construct of a particular time, a response to a particular strategic situation, and a solution to a set of particular operational problems. The emergence of doctrine was the result of the confluence of a number of specific mid-nineteenth century political, economic, social and military factors that reshaped Europe's cultural fabric and altered the balance of power between the continental states in favour of the Kingdom of Prussia. In fact, doctrine was a product of the Prussian General Staff System and emerged under the direction of one of its most gifted leaders, Field Marshal Helmuth von Moltke.

The strategic, operational and tactical problems facing today's military professionals and defence thinkers bear little resemblance to those with which Moltke and his subordinates contended 150 years ago. Since the emergence of doctrine, the art of war has witnessed the Mechanisation and Combined Arms Revolutions in Military Affairs, the onset of the Atomic Age, the rise and collapse of the Soviet Union, the emergence of a US-dominated unipolar world, and the birth of Osama bin Laden-inspired terrorism. The onset of the Information Revolution in Military Affairs adds a further dimension to an already extremely complex strategic and operational environment. The result is that modern warfare is characterised by

> complexity, ambiguity, uncertainty, and volatility, and by the fact that they all tend to be 'wicked problems'—problems that are intractable and circular with complex inter-dependencies, and where solving one part of the problem can create further problems, or make the whole problem greater.[2]

While it is true that war, in essence, remains 'an act of force to compel our enemy to do our will',[3] it is now waged at a scale, cost, tempo and complexity that is well outside the experience and knowledge base of commanders, staffs and thinkers of Moltke's era. Yet, despite warfare's evolution beyond the context of its creation, doctrine has endured as the intellectual cornerstone of military organisations.

From Moltke to bin Laden questions the continued relevance of doctrine in light of the warfighting challenges that today's military professionals must face. The question posed in the pages of this study paper is whether it is now time for armed forces to consign doctrine to the past and replace it with a different intellectual framework more suited to the challenges of the twenty-first century. In the course of its analysis, this paper will assert that doctrine has not ruled with the supremacy that its advocates might contend. In fact, there was one major power exception to the universal acceptance of the necessity of institutional indoctrination—namely the British Army before, during and after the First World War. This paper will establish that despite Britain's participation in numerous diverse operations ranging from small tribal skirmishes to a global-scale conflict against conventionally and unconventionally armed, structured and trained opponents, its military leaders negotiated the problems and challenges of these actions without recourse to a doctrine. This was because the British Army employed a different intellectual mechanism that more closely

---

2 John Kiszely, *Post-Modern Challenges for Modern Warriors*, Defence Academy of the United Kingdom, The Shrivenham Papers No. 5, Shrivenham, 2007, p. 8.

3 Carl Von Clausewitz, *On War*, Michael Howard and Peter Paret (eds and trans), Princeton University Press, Princeton, 1989, p. 75.

resembled an 'institutional ethos'. While this paper does not identify doctrine's replacement, it does stress the need for military organisations to seek an alternative, modern, intellectual structure, one with greater applicability to the rapidly evolving and complex nature of the wars of the present and the future.

From Moltke to bin Laden will build its argument in a series of stages. In its initial stages it will establish the role and importance of doctrine within modern military organisations and describe the context of doctrine's origin in the nineteenth century. It will then consider the circumstances that led to the British Army's consideration and subsequent rejection of the need for a doctrine, and will highlight the transformation of the British Army on the Western Front in the First World War as an example of a non doctrine-orientated adaptation system. The paper will then explain how institutional ethos operated within the British Army as an alternative learning mechanism to doctrine. In its final stage, *From Moltke to bin Laden* contrasts the extreme complexity of the contemporary warfighting environment with the art of war as it was pursued in the nineteenth century and, in so doing, highlights the requirement for a new intellectual framework capable of supplanting doctrine.

The nature and importance of doctrine

Contemporary military leaders invest enormous resources in the writing, dissemination and inculcation of doctrine. Their willingness to make such a sizeable investment is born of the belief that doctrine plays a pivotal role in determining the effectiveness of the forces under their command. In war, the stakes are indeed high, and embarking on a military campaign with a flawed or inferior doctrine can have a disastrous effect on the fate of an army—and even a nation. The German Army's crushing defeat of the French Army in May 1940 provides a compelling example of the consequences of poorly conceived doctrine.[4] The French interpretation of the lessons of the First World War prompted them to implement a mode of warfighting that depended on the waging of a highly controlled, methodical, set-piece battle in which artillery was the dominant arm. Unfortunately for France, its military planners and theorists based their conclusions on a narrow analysis of their force's experiences during the period of stalemate on the Western Front. In so doing, they failed to take into account the possibilities of mechanisation and the advances in combined arms and mobile warfare tactics that emerged in the conflict's final year, and which resulted in the restoration of a more fluid battlespace. The German Army did not commit this error. Its leaders inculcated their forces with a doctrine that integrated mobility, mechanisation, the use of multiple arms and mission flexibility into a mode of attack that it used to overwhelm first Poland, then Denmark and Norway, and finally France and the Low Countries in a series of lightning campaigns, each lasting just weeks, and which also ousted the British Expeditionary Force from the continent.

Military practitioners and defence thinkers identify doctrine manuals as the capstone documents for the shaping of their force's warfighting capability. Senior officers and defence thinkers alike have acknowledged doctrine's role in defining a military organisation's intellectual focus. Michael Evans, for example, describes doctrine as of 'critical importance in a modern army; it helps provide a philosophical

4 See Robert Allan Doughty, *The Seeds of Disaster: The Development of French Army Doctrine, 1919-1939*, Archon Books, Hamden, 1985.

impetus for thinking about the needs of learning, anticipation and adaptation by attempting to identify the constants and the variables in war'.[5] It is, according to Evans, 'the foundation of military knowledge'.[6]

No current military organisation places greater faith in the utility of doctrine than the United States (US) defence establishment. For the US military, particularly the US Army, doctrine is the bedrock upon which its intellectual foundation is laid. In 1992 its Chief of Staff, General Gordon R Sullivan, highlighted doctrine's importance in the lead-up to the force's Gulf War review. He identified doctrine as the 'catalyst for change across the army'. Sullivan continued:

> doctrine provides the framework for institutional changes within the Army—changes to the structure of our organization, to training and leader development programs and to the equipment we develop and procure.[7]

Sullivan expected doctrine to serve as the force's guide on its journey towards the army of the future.[8] In the aftermath of the American trilogy of successes in the First Gulf War, Panama and the Cold War, the Commander of US Army Training and Doctrine Command, General Frederick M Franks, Jr, wrote that more than ever before 'doctrine will help us keep ahead of change, as well as reverse the downturn in effectiveness that normally follows battlefield victory'. Doctrine writing, Franks believes, would be the method by which the army defined its requirements for the future.[9] Another US general, Donn A Starry, elevated doctrine to the highest of laws, writing that it is the 'first and great commandment'.[10] US defence thinker Roger J

---

5 Michael Evans, *Forward from the Past: The Development of Australian Doctrine 1972 – Present*, Land Warfare Studies Centre, Canberra, 1999, p. 4.

6 Ibid., p. 2.

7 Gordon R Sullivan, 'Doctrine: A Guide to the Future', *Military Review*, Vol. LXXII, No. 2, February 1992, p. 4.

8 Ibid., p. 9.

9 Frederick M Franks, Jr, 'Continuity and Change: Discussing Our Evolving Doctrine', *Military Review*, Vol. LXXI, No. 10, October 1991, p. 1.

10 Donn A Starry, 'Changing an Army: An American Perspective' in Michael Evans (ed), *Changing the Army: The Roles of Doctrine, Development and Training*, Land Warfare Studies Centre, Duntroon, 2000, p. 12.

Spiller has rated doctrine in second or third place in a modern army's hierarchy of professional concerns.[11] Nor is this adherence to doctrine a recent development. The US Army wedded itself to doctrine shortly after the end of the First World War and has only deepened that association ever since. In fact, doctrine is now regarded as fundamental to the development of all aspects of American military power including administration, education, training, weapon design, procurement and the concept of waging war.

Despite this depth of belief in doctrine's critical importance, there is little agreement over what precisely constitutes 'doctrine'. Scholars and military practitioners approach doctrine largely on the intuitive level, and there is no single approved definition. In practice, definitions follow institutional or personal whim and vary between individuals, services, nationalities and time periods. A useful generic definition is provided by *Brassey's Encyclopedia of Land Forces and Warfare* which offers: 'doctrine provides a common reference point, language, and purpose, uniting the actions of many diverse elements into a team effort'.[12]

As the *Brassey's* definition suggests, unity is one of the outcomes credited to doctrine. Military organisations employ doctrine to provide their institutions with a commonality of purpose through the establishment of shared interpretive frameworks and operational procedures. At its most effective, doctrine enables the members of a military organisation to work towards common goals by universally accepted and understood means. The interwar-era Soviet general and military theorist, General Alexandr A Svechin, argued that doctrine must exist 'so that in the realm of military thought an army does not represent human dust, but a cohesive whole'.[13] The British general and theorist J F C Fuller described the goal of doctrine as providing 'that spirit of unity so essential to success'.[14] An early US thinker and a future Commandant of the

---

11 Roger J Spiller, 'In the Shadow of the Dragon: Doctrine and the US Army after Vietnam' in Jeffrey Grey and Peter Dennis (eds), *From Past to Future: The Australian Experience of Land/Air Operations*, Australian Defence Force Academy, Canberra, 1995, p. 6.

12 Franklin D Margiotta, *Brassey's Encyclopedia of Land Forces and Warfare*, Brassey's, Washington DC, 2000, p. 294.

13 Quoted in Charles Grant, 'The Use of History in the Development of Contemporary Doctrine' in John Gooch (ed), *The Origins of Contemporary Doctrine*, Strategic & Combat Studies Institute, Camberley, 1997, p. 7.

14 J F C Fuller, *Training Soldiers for War*, Hugh Rees Ltd, London, 1914, p. 43.

USMC, Major John H Russell wrote that doctrine 'means a teaching that provides for a mutual understanding among the commissioned personnel of a military organization. In plain words "teamwork".'[15] Scholars have also recognised doctrine's unifying role. Jack Snyder asserts that doctrine provides 'a criterion for the establishment of standard operating procedures ... [and] helps to provide a simple, coherent, standardized structure'.[16] Russell A Hart summarises doctrine as 'the intellectual substance that binds together an army's organization, equipment, personnel, training, tactics, and provides the army's basic procedures'.[17]

Doctrine's establishment of 'uniformity in mental processes'[18] further benefits a military organisation by improving its ability to cope with complexity and reduce the uncertainty of war. By developing techniques for scanning and organising information, and by authorising mechanisms for structuring solutions, doctrine allows an army to streamline adaptations to novel challenges or changes in the nature of war and to identify those innovations required to achieve a major transformation in capability.[19]

Clearly then, military organisations do not achieve indoctrination by chance. For a force to inculcate the tenets it holds true, doctrine must be recorded, disseminated and accepted throughout the organisation. Without the successful recording, dissemination and acceptance of these doctrinal tenets, a military organisation's indoctrination does not occur, nor is the presumed doctrine a valid representation of the institution's intellectual intent, irrespective of the claims of its promoters. Moreover, the development of doctrine requires central control, direction and support from the highest levels, and

---

15 John H Russell, 'A Plea for a Mission and a Doctrine', *The Marine Corps Gazette*, Vol. I, No. 2, June 1916, p. 115.

16 Jack Snyder, *The Ideology of the Offensive: Military Decision Making and the Disasters of 1914*, Cornell University Press, Ithaca, 1984, p. 27.

17 Russell A Hart, *Clash of Army: How the Allies Won in Normandy*, Lynne Rienner, Boulder, 2001, p. 170.

18 T H Low, 'Discussion on a Plea for a Mission and a Doctrine', *The Marine Corps Gazette*, Vol. I, No. 3, September 1916, p. 332.

19 Snyder, *The Ideology of the Offensive*, p. 27.

it must have the army's official stamp of approval.[20] The Australian Army's *LWD-1* notes that, for doctrine to be effective, it must be 'embraced by its practitioners'.[21]

The perceived need for doctrine is now so deeply embedded in the psyche of contemporary military organisations that any inherent liabilities in the concept's adoption are usually ignored or overlooked. Yet risks in implementing doctrine do exist, and some undesirable features of an organisation's indoctrination can be extremely debilitating, if not dangerous.

Perhaps doctrine's most risky side-effect is, as Fuller wrote, 'that it is apt to ossify into a dogma'. He continued, 'and be seized upon by mental emasculates who ... are only too grateful to rest assured that their actions, however inept, find justification in a book'. To highlight the danger, Fuller equated adherence to dogma with 'self destruction or suicide through inertia of [the] mind'.[22] Another assessment cautioned that 'doctrine that spurns new information in order to preserve the old order for its own sake is not doctrine, it is dogma'.[23] Richard Overy observes that the twentieth century is not lacking in examples of doctrines that solidified like 'a slowly moving lava flow'.[24]

British officers first voiced concerns over the possibility of doctrine becoming dogma during the years of army reform following the Anglo-Boer War of 1899-1902.[25] This wariness remained strikingly apparent for almost the entire twentieth century. In 1993, when Rear Admiral J H S McAnally undertook the preparation of the Royal Navy's *The Fundamentals of Maritime Doctrine*,[26] one of the objections he had to

---

20 Evans makes this point in *Forward from the Past*, p. 69. See also James R McDonough, 'Building the New FM 100-5: Process and Product', *Military Review*, Vol. LXXI, No. 10, October 1991, p. 4.

21 Australian Army, *The Fundamentals of Land Warfare*, Department of Defence, Canberra, 1988, p. ii.

22 J F C Fuller, *The Foundations of the Science of War*, Hutchinson & Co, London, 1925, p. 254.

23 McDonough, 'Building the New FM 100-5', p. 5.

24 Richard Overy, 'Doctrine Not Dogma: Lessons from the Past' in Michael Duffy, Theo Farrell and Geoffrey Sloan (eds), *Doctrine and Military Effectiveness: Proceedings of the Conference held at the Britannia Royal Navy College, January 16th-17th, 1997*, Strategic Policy Studies Group of the Britannia Royal Naval College & Exeter University, Exeter, 1997, p. 34.

25 For an example see *Report of a Conference of General Staff Officers at the Staff College, 9th to 12th January, 1911*, HMSO, London, 1911, pp. 5-9.

26 See *The Fundamentals of British Maritime Doctrine*, 1996.

overcome was the concern that doctrine would impede thinking and bind officers to inflexible rules.[27]

It is largely due to the fear of dogma that those who write doctrine insist that its creation must be an ongoing activity. Dynamic military organisations strive to revise their doctrines as the potential for war evolves. To this end a former British Chief of the General Staff, Field Marshal John L Chapple, insisted that doctrine was not a set of rules or prescription for success.[28] Moreover, the British Army's first official definition of the term contained the qualifying statement that doctrine 'is authoritative but requires judgement in application'.[29] Fuller concluded that to be effective doctrine must also be 'elastic'.[30]

Doctrine's institutional sanction is also necessary in order to prevent another risk-laden side-effect: the development of sub-doctrines or variants. These are doctrinal ideas that originate either from individuals or from a military organisation's minor centres of intellectual development. If unchecked, these irregular concepts can prevent a military organisation from adopting a unified conception of war, or possibly even undermine its official writings. In any consideration of doctrine, it is necessary to distinguish between the ideas of the individual and that of the institution. No matter how brilliant or admired a thinker's work may be, it does not represent doctrine if left unsanctioned by the institution.[31]

A further negative outgrowth of doctrine is its ability to encourage stereotyped thinking. The cost of an imposed uniformity of outlook is the closure of options, in particular to unanticipated or non-standard problems. In 1902, for example, the highly influential British military thinker and educator G F R Henderson summarised the British Army's attitude towards stereotyped thinking. He observed that manuals did not bind the great leaders of the past. Commanders such as Napoleon or Stonewall Jackson 'would have no fetters cramping their intelligence and common-sense'. The key to success in battle, Henderson noted, was 'the habit of using the wits, of

27 J H S McAnally, 'The Purposes and Benefits of Doctrine: Why go to all the Trouble of Having One?' in Michael Duffy, et al, (eds), *Doctrine and Military Effectiveness*, p. 7.

28 Great Britain, General Staff, *Design for Military Operations – The British Military Doctrine*, 1989, p. vii.

29 Ibid., p. 3.

30 Fuller, *The Foundations of the Science of War*, p. 254.

31 Ibid. See also Evans, *Forward from the Past*, p. 4.

subordinating the rules of theory to the needs of the moment, and if necessary discarding them *in toto*'. Moreover, he observed that, once a commander has become accustomed to stereotyped proceedings 'the danger is great'.[32]

The oft-added warning that appears in the introductions to doctrinal manuals or to definitions of the term itself, that 'doctrine is authoritative but not prescriptive',[33] underscores the fear that pervades military institutions that commanders will seek automatic solutions from manuals rather than utilising applied thought. There would be no need for this clarification if an institution's recourse to doctrine did not signal the possibility of stereotyped thinking.

While this study paper will not offer a single definition for doctrine from amongst the vast array of variations, it is possible to distil those features that serve to declare its existence. These are the essential common denominators by which an observer can determine whether a military organisation has achieved indoctrination. When such conditions are not met, claims for the existence of doctrine are baseless. However, it also remains true that the presence of these traits alone does not guarantee that a force's doctrine will be appropriate and effective; the only real test of this is war.

Indoctrination is typified by two primary characterisations. First, doctrine is an intellectual construct that a military organisation employs to define its concept of war, and by which it makes changes in its method of waging war. Second, doctrine must be uniform, embraced by the entire institution, and thoroughly inculcated at all levels of the organisation. It is more than the publication of manuals—it must go beyond the printing press. Moreover, these two primary characteristics suggest the existence of a further four secondary characteristics, namely: third, doctrine is a mechanism for the learning and dissemination of lessons. Fourth, doctrine is recorded in written form so as to facilitate its distribution and application. Fifth, doctrine is necessarily dynamic; otherwise it will ossify into dogma. Sixth, doctrine is the expression of a military institution's officially sanctioned ideas. It is not the property of an individual, nor can it exist outside an organisation's corporate mentality.

---

32 'Tactics of the Three Arms Combined', reprinted in G F R Henderson, *The Science of War: A Collection of Essay and Lectures 1891-1903*, Longmans, Green and Co., London, 1910, pp. 71-72.

33 Department of the Army, *FM 3-0 Operations*, Washington DC, 2001, pp. 1-14. The Australian Army's capstone doctrinal manual employs the phrase 'It is authoritative, but requires judgement in its application'. See Australian Army, *Land Warfare Doctrine 1 – Fundamentals of Land Warfare*, Land Warfare Development Centre, Puckapunyal, 2002, p. 5.

These observations illustrate that doctrine fulfils a number of key functions within a military organisation. It guides thought, articulates innovation and adaptation, unifies a force with a commonality of training and objective, and acts as a common belief system shared by the institution's members. Within this process it also offers both benefits and risks. However, these attributes are representative of what doctrine *does* rather than what it actually *is*. Stripped of its various embellishments, doctrine is a tool, or perhaps an intellectual device, whose purpose is to facilitate war planning and generate combat effectiveness.

The origins of doctrine

The development of doctrine is associated with the work of Field Marshal Helmuth von Moltke during his tenure as Chief of Staff from 1857 to 1888, first of the Prussian General Staff and, after 1871, the Great General Staff. Moltke, through his control of the General Staff, articulated, disseminated and inculcated a doctrine of warfighting that imbued the Prussian Army with heightened effectiveness and capability. In so doing, Moltke ushered in the modern era of military command, control and management.

A few scholars assert that some military organisations also possessed articulated ideas on the conduct of war that either were or represented the function of doctrine, in advance of the Moltkean Prussian Army. However, it would be misleading to associate the warfighting methods of, for example, an Alexander, a Napoleon or a Frederick with doctrine as it functions in present-day military organisations. The 'doctrines' of these figures, even when they chose to commit their thoughts to paper, were nothing more than the fleeting personal choices of commanders, if great ones. Their ideas were the product of transient forceful leadership and did not represent institutionally inculcated beliefs.[34] Following a similar line of reasoning, the identification of drill regulations as examples of early forms of doctrine is equally false.[35]

Although they often carry the label of 'doctrine', procedural manuals are not the subject of this study paper, and their existence is not proof of an institution's possession of a doctrine. This study paper's focus remains on the higher level philosophical form of doctrine and, from this perspective, Moltke was the first military leader to articulate a concept of war on a systematic, professional and ongoing basis. For this reason he deserves credit as its originator.

---

34 Paul H Herbert, *Deciding What Has to Be Done: General William E DePuy and the 1976 Edition of FM 100-5, Operations*, Combat Studies Institute, Fort Leavenworth, 1988, p. 3; and Jay Luvaas, 'Some Vagrant Thoughts on Doctrine', *Military Review*, March 1986, pp. 56-57.

35 See William O Odom, *After the Trenches: The Transformation of US Army Doctrine, 1918-1939*, Texas A&M University Press, College Station, 1999, pp. 5-6.

Moltke was a military intellectual and reformer who sought improvements in the Prussian Army's management, administration, organisation and education as a means to increase the force's effectiveness. Victories over Denmark (1864), Austria (1866) and France (1870-71) which culminated in German unification, underscored the overwhelming advantages Moltke's leadership had provided the Prussian Army over its rivals.

Moltke's decision-making processes and transformational objectives were a product of his age. Shaping his interpretative framework was Prussia's security situation and geography as well as the effects of two Revolutions in Military Affairs. Prussia was one of the many German states—albeit a leading one—that resonated with the desire for Germanic unification on its own terms. Prussia's geographic location in central Europe, surrounded by stronger great power rivals, also played a significant part in shaping Moltke's thought. He was immersed in the social and political legacies of the French Revolution and the rapid technological changes introduced by the Industrial Revolution. The interaction of these strategic and transformational factors created a multidimensional security situation of challenging complexity.

The most pressing strategic question facing the German states in the nineteenth century was the issue of unification. At the century's midpoint the German people remained divided into a number of independent states of which Austria and Prussia were the most powerful. The Germans faced two questions: first, whether or not to seek unification and thereby move towards world power status on par with Britain or France, or remain a league of provincial-sized states united only by language and custom. The second question involved which state around which to unify: Prussia or Austria. These options dictated the position and role that Prussia would play in central Europe and beyond.

Further complicating Prussia's strategic decision-making was its geographic location and, in particular, the lack of strong geographic features upon which to base a defence. Prussia had potential rivals on all sides: Russia to the east, Austria to the south, France to the west and, to the north, a declining Sweden. Only the Baltic Sea protected Prussia's northern approaches, while the lack of other natural boundaries laid the country open to invasion, particularly from the east. Geography and rivalry conspired to forced German planners to confront an extremely difficult security issue—the challenge of a two-front war.

While the French Revolution in Military Affairs predated Moltke's period of service, its effect on the nature of war remained starkly evident during his time. In 1806, Prussia's crushing defeat at Napoleon's hands at the twin battles of Jena-Auerstadt revealed the inferiority of an unreformed Frederickian-era army when challenged by one invigorated and transformed by the liberty, energy and national purpose of Revolutionary France. The effect of the *levée en masse* was to provide the armies of the Revolution and Napoleon with enormous manpower resources that Europe's monarchical states could not initially match, and the removal of the *ancien régime* opened the French Army's leadership ranks to the talents of an entire population rather than just the members of a small ennobled caste.

Under the guidance of General Gerhard von Scharnhorst, Prussia responded to the French Revolution in Military Affairs by changing the basis of membership in the army from a long-service regular force to one centred on short-term conscription followed by a reserve service obligation. In effect, Scharnhorst reconstituted the force from one representative of the age of despotism into one that could serve as the model for a national army.[36] Parallel to these structural modifications was the professionalisation of the officer corps which transformed into a body in which knowledge, education and experience became the means of advancement. To support this goal Prussia founded a number of new schools and emphasised the necessity for ongoing professional military education. The result of these initiatives was the creation of a well trained and led standing army, supported by a large expansion base of experienced soldiers, that could generate a force of sufficient size to counter Prussia's enemies—but not so large as to be a burden on the state's limited resources.[37] Moltke oversaw further improvements in Prussia's military recruitment and education systems once he became Chief of Staff. These resulted in a doubling of the size of the field army and a further strengthening of its reserve capability.[38]

---

36 Hajo Holborn, 'The Prusso-German School: Moltke and the Rise of the General Staff' in Peter Paret (ed), *Makers of Modern Strategy from Machiavelli to the Nuclear Age*, Princeton University Press, Princeton, 1986, p. 282.

37 Archer Jones, *The Art of War in the Western World*, Oxford University Press, Oxford, 1989, pp. 393-94; and Arden Bucholz, *Moltke, Schlieffen and Prussian War Planning*, Berg, New York, 1991, p. 22.

38 Arden Bucholz, *Moltke and the German Wars, 1864-1871*, Palgrave, Houndmills Basingstoke, 2001, p. 74.

While it is evident that Moltke built on a foundation laid by his predecessors in responding to the legacy of the French Revolution, the Prussian Army's adaptation to industrialisation, however, was largely due to his influence alone. Dennis E Showalter in *Railroads and Rifles: Soldiers, Technology, and the Unification of Germany* observes that the Industrial Revolution ushered in an era of technological change from which the armies of the period were by no means exempt.[39] Showalter highlights the enormous influence on strategy and operations of the development of the railroad and telegraph as a means to mobilise, organise, move, concentrate and sustain large bodies of troops. Equally significant was the effect on tactics of the engineering breakthroughs and improved casting and metallurgy techniques associated with the appearance of breech-loading rifles and steel cannons.[40]

Moltke's concept of war involved combining 'rapid mobilization, transportation, deployment, movement, and combat into one continuous sequence, making every effort to bring superior numbers to bear in the final decisive battle'.[41] Victory through battle against the armies of Prussia's European rivals was his sole objective. Moltke used the General Staff to perfect this method of waging war. In late 1857 when he assumed the position of the Chief of the General Staff, he commanded a minor agency that was subordinated to the War Ministry and which lacked institutional credibility. As one historian comments, it had 'little reputation, real or imagined'.[42] Moltke's success, however, was such that within a decade he had raised the General Staff to a position of dominance within the army and institutionalised under his control the processes with which to prosecute modern war. He crafted the staff into an instrument that combined 'flexibility and initiative at the local level with conformity to a common operational doctrine and to the intentions of the high command'.[43] In so doing he

39 Dennis E Showalter, *Railroads and Rifles: Soldiers, Technology and the Unification of Germany*, Archon Books, Hamden, 1975, p. 11.

40 A summary of the age's military technology advances can be found in Dennis E Showalter, 'The Prusso-German RMA, 1840-1871' in Macgregor Knox and Williamson Murray (eds), *The Dynamics of Military Revolution 1300-2050*, Cambridge University Press, Cambridge, 2001, pp. 92-113.

41 Gunther E Rothenberg, 'Forward' in Daniel J Hughes (ed), Daniel J Hughes and Harry Bell (trans), *Moltke on the Art of War: Selected Writings*, Presidio Press, Novato, 1993, p. viii.

42 Bucholz, Moltke, *Schlieffen and Prussian War Planning*, p. 39.

43 Gunther E Rothenberg, 'Moltke, Schlieffen, and the Doctrine of Strategic Envelopment' in Paret (ed), *Makers of Modern Strategy*, p. 301.

endowed the Prussian Army with a staff corps whose individual members would provide approximately the same solution to any given situation.[44] Subsequent thinkers recognised the link between the function of the General Staff and the development of doctrine. Russell, the US Marine Corps' doctrine advocate, believes that doctrine writing was not the task of a single officer but rather a labour for the General Staff as a whole.[45] The General Staff, to borrow Spenser Wilkinson's words, had become 'The Brain of an Army'.[46]

The Prussian General Staff was responsible for developing plans for the organisation and mobilisation of the army, and for its peacetime training and education.[47] Ultimately, its function was to plan and direct future war.[48] The Prussian General Staff Officer, unlike those of other armies, was more than simply a technical adviser. Staff officers were not only duty bound to offer their advice, it was their prerogative—and, significantly, they also had the right to issue orders in their commander's name.[49]

The General Staff controlled the Prussian Army's strategic, operational and tactical war planning, educational institutions, historical studies, and key training activities such as staff rides and war games. Each of these tasks became the means for the dissemination and inculcation of Moltke's ideas. For example, the General Staff oversaw the publication of *Military Week* (*Militär-Wochenblatt*), a journal that featured articles on war history. The staff also sponsored the publication of books, including a series of historical studies on Prussia's wars. Moltke himself was a prolific author and a frequent contributor to *Military Week*, which published some of his better known essays, including 'Instructions for Large Unit Commanders' which remained the foundation of German formation operations for seventy years.[50]

Once a young officer completed his staff training he effectively became a vehicle for the distribution of Moltke's concept of war. Moltke handpicked the members of the General Staff directly from the graduating class of the War Academy, the world's first

---

44 Ibid.

45 Russell, 'A Plea for a Mission and a Doctrine', p. 116.

46 See Spencer Wilkinson, *The Brain of an Army*, Greg Revivals, Aldershot, 1992.

47 Holborn, 'The Prusso-German School' in Paret (ed.), *Makers of Modern Strategy*, p. 283.

48 Bucholz, *Moltke and the German Wars: 1864-1871*, p. 17.

49 Ibid., p. 59.

50 Ibid., p. 63.

modern school of higher professional military education. He selected only the best prospects, and discarded to regimental duties those who did not meet his exacting standards. Typically, new staff officers first served in the typographical department, the section in which Moltke himself had commenced his stellar career. Following the German practice, all staff officers received appointments to field formations at set points in their careers. Throughout their service, Moltke's disciples alternated between staff and field appointments, thereby assuring that his ideas circulated throughout the army.[51]

Moltke was not a theoretician, but rather a pragmatist whose ideas and writings focused on how most efficiently to destroy the enemy's will to resist. His processes embodied 'a practical method of how to do this using the technological and organizational tools of the mid-nineteenth century'.[52] By the time Moltke became the Chief of the General Staff, Prussia was a great power—but only just. His genius lay in his ability to define and disseminate throughout the Prussian Army—through the agency of the General Staff—a concept of war that took advantage of the strengths of a national army, helped solve the country's strategic problems and allowed Prussia to identify and exploit the military applications of the emerging technologies of the Industrial Revolution in Military Affairs. Moltke rationalised the complexities of his age by institutionalising a self-replicating system of interpretation and action that contemporary military professionals would identify as the first articulated doctrine of war. In effect, Moltke created a General Staff with a single vision of war that 'acted as a nervous system animating the lumbering body of an army, making possible the articulation and flexibility which alone rendered it an effective military force'.[53]

Moltke's achievements were far-reaching. The military and political results of his innovations were starkly apparent in the Prussian Army's relatively easy victories over its opponents and the creation of a unified Germany dominated by Prussia. As one historian asserts with no hint of exaggeration, the rise of the General Staff was the most important military advancement of the nineteenth century.[54]

---

51 Bucholz, Moltke, *Schlieffen and Prussian War Planning*, p. 49; and Michael Howard, *The Franco-Prussian War: The German Invasion of France, 1870-1871*, Collins, London, 1967, p. 25.

52 Bucholz, *Moltke and the German Wars: 1864-1871*, p. 54.

53 Howard, *The Franco-Prussian War*, p. 24.

54 John A English, *Marching Through Chaos: The Descent of Armies in Theory and Practice*, Praeger, Westport, 1996, p. 50.

The British Army's rejection of doctrine

British officers of the First World War era understood the purpose of doctrine, and they knew that their continental rivals had developed and inculcated doctrines for their own organisations. Yet, despite this clarity of vision, the British Army did not become indoctrinated until relatively recently. It was not until 1989, with the publication of *Design for Military Operations – The British Military Doctrine*, that the British Army accepted as an institution that doctrine had a worthwhile role to play in its military actions. That the British Army was the last major Western land force to accept the necessity for doctrine was not the result of mischance, ignorance or lassitude, nor was it an oversight or error. It was a deliberate calculation born of the interaction of rational decision-making and the institution's own cultural bias. That the British Army was so late to accept indoctrination was its own carefully considered choice.

At first glance this may appear an unsupportable conclusion given the numerous references to doctrine in the literature of the British Army. For example, almost seventy years before the publication of *Design for Military Operations*, the first sentence of the 1920 edition of the British Army's *Field Service Regulations* assumes indoctrination when it states, 'the Army will be trained in peace and led in war in accordance with the doctrine contained in this volume'.[55] The army employed similar language in the manual's subsequent printings until the major 1935 revision when the sentence quoted above, as well as the term 'doctrine', vanished without explanation.[56] Many military thinkers of the period also mentioned doctrine in their publications. Fuller, for example, in his 1914 book, *Training Soldiers for War*, included a six-page discussion on doctrine.[57] Contemporary scholars have freely participated in fostering the illusion of British Army indoctrination by incorrectly referring to its existence in

55 General Staff, War Office, *Field Service Regulations, Volume II, Operations (Provisional)*, HMSO, London, 1920, p. 13.

56 See General Staff, War Office, *Field Service Regulations, Volume II, Operations – General*, HMSO, London, 1935.

57 Fuller, *Training Soldiers for War*, pp. 41-46.

their writing, such as in the title of an essay by Brian Bond, 'Doctrine and Training in the British Cavalry, 1870-1914'.[58]

Interestingly, while most recent scholarship accepts that the present British Army now has a doctrine, there has been no agreement on when the force accepted indoctrination.[59] Brian Holden Reid states authoritatively that the British Army lacked a coherent doctrinal philosophy throughout the twentieth century.[60] In contrast, John Stone offers the opinion that the British Army possessed an articulated body of doctrine as early as 1909.[61] With such divergence in views, it is not surprising that historians have argued for and against a variety of dates marking the moment the British Army accepted the need for a doctrine and acted on its implementation. Scholars have nominated as the pivotal point such events as the publication of the first edition of the *Field Service Regulations* in 1909,[62] the issuing of colonel and armour theorist Charles Broad's 1929 manual *Mechanized and Armoured Formations*,[63] and General Bernard Montgomery's efforts to standardise the 8th Army's procedures in North Africa during the Second World War.[64] The army itself has selected 1989, the year in which it released *Design for Military Operations – The British Military Doctrine*, as the point that it became indoctrinated for the first time.[65]

The British Army's thinkers began to question the need for doctrine shortly after the end of the war in South Africa in 1902. Britain's eventual victory over the Boer Republics did not mitigate the army's poor initial performance, which led to a

---

58 Brian Bond, 'Doctrine and Training in the British Cavalry, 1870-1914' in Michael Howard (ed), *The Theory and Practice of War: Essays Presented to Captain B H Liddell Hart*, Oxford, London, 1965, pp. 95-125.

59 For example, see Hew Strachan, 'The British Way in Warfare' in David Chandler (ed), *The Oxford Illustrated History of the British Army*, Oxford University Press, Oxford, 1994, p. 430.

60 Brian Holden Reid, *A Doctrinal Perspective, 1988-98*, Strategic & Combat Studies Institute, Camberley, 1998, p. 12.

61 John Stone, 'The British Army and the Tank' in Theo Farrell and Terry Terriff (eds), *The Sources of Military Change: Culture, Politics, Technology*, Lynne Rienner Publishers, Boulder, 2002, p. 191.

62 See General Staff, War Office, *Field Service Regulations, Part I, Operations,* HMSO, London, 1909.

63 See General Staff, War Office, *Mechanized and Armoured Formations*, HMSO, London, 1929.

64 See John A English, *The Canadian Army and the Normandy Campaign: A Study in Failure in High Command*, Praeger, Westport, 1991, p. 160.

65 See Great Britain, General Staff, *Design for Military Operations*.

series of defence reforms.[66] These reforms did not extend as far as the inculcation of a doctrine, however, and a closer examination of the issue points to the conclusion that the British Army simply refused to indoctrinate itself. Hindering the army's adoption of doctrine were four structural and cultural factors including, significantly, the fact that the army lacked a suitable general staff structure which would allow the development of doctrine. The British Army also adhered to the belief that doctrine was an inappropriate mechanism for a military organisation that had a multitude of greatly differing responsibilities across a global empire. In addition, the army lacked a concept of war around which to shape a doctrine and also harboured fears that doctrine would lead to stereotyped thinking and degrade the role of commanders in battle. This study paper will examine each of these factors in turn and, in so doing, demonstrate that their combined effect was to render it impossible for the British Army to indoctrinate itself.

The British Army's lack of an appropriate general staff structure for the development of a doctrine

The Prussian General Staff was the essential conduit through which Moltke outlined and institutionalised a doctrine for the Prussian Army. Moltke achieved this because he commanded a strong, durable organisation charged with guiding the development and dissemination of Prussia's concept of war. Without meeting these preconditions it is unlikely that a military institution would be able either to agree upon a doctrine or to inculcate it throughout the organisation. An examination of the British Army's inability to advance the idea of doctrine reveals that these are precisely the characteristics that it lacked.

The First World War-era British Army lacked the essential strong, institutional leadership that could have served as a rallying point for the advancement of a doctrine. Technically, after 1904, the army did not even have a commander. In that year the government abolished the position of Commander-in-Chief and replaced the force's single head with a committee known as the Army Council. Within the ranks of the Army Council were four military members: Chief of the General Staff, Adjutant-General, Quarter-Master General, and Master-General of the Ordnance. A civilian finance member also sat on the council, which was chaired by the Secretary of State

66 For the story of these reforms, see Edward M Spiers, *Haldane: An Army Reformer*, Edinburg University Press, Edinburg, 1980.

for War. In 1909 the title 'Chief of the General Staff' was amended to 'Chief of the Imperial General Staff'.[67]

While the most influential of the Army Council's military members, the Chief of the Imperial General Staff was merely one element of a committee, with powers and responsibilities equivalent to those of the other members. In fact, each military member had his own right of access to the Secretary of State for War, and was solely responsible for the affairs of his department. The stultifying effect of operating as part of a committee meant that no senior officer had responsibility for the force's intellectual development—a critical requirement for a military organisation's indoctrination.[68]

Of the four military positions, the office of the Chief of the Imperial General Staff was the most suited to serve as the wellspring of doctrine. The role's focus was operations, training and intelligence, and the Chief of the Imperial General Staff performed these tasks through the General Staff. Yet, the General Staff was itself a recent creation.[69] The functions of the General Staff were:

> [to] advise on the strategical distribution of the Army; to supervise the education of officers and the training and preparation of the Army for War; to study military schemes, offensive and defensive; to collect and collate military intelligence; and, finally, to direct the general policy in Army matters and to secure continuity of action in the execution of that policy.[70]

The function and status of the British General Staff compared poorly, however, with that enjoyed by the German General Staff which had a mandate for the preparation of doctrine for the German Army.[71] Moreover, unlike the staffs of the German and French

---

67 Shelford Bidwell and Dominick Graham, *Fire-Power: British Army Weapons and Theories of War 1904-1945*, George Allen & Unwin, London, 1982, p. 46; and Halik Kochanski, 'Planning for War in the Final Years of *Pax Britannica*, 1889-1903' in David French and Brian Holden Reid (eds), *The British General Staff: Reform and Innovation c.1890-1939*, Frank Cass, London, 2002, p. 25.

68 Harold Winton, *To Change an Army: General Sir John Burnett-Stuart and British Armored Doctrine, 1927-1938*, University Press of Kansas, Lawrence, 1988, p. 190.

69 The organisation of the General Staff is described in Brian Bond, *The Victorian Army and the Staff College, 1854-1914*, Eyre Methuen, London, 1972, pp. 212-43.

70 Ibid., p. 18.

71 See Wilkinson, *The Brain of an Army*.

armies, the British General Staff lacked its own identity and spirit, and failed to provide the officer corps with an army-wide body with which its members could identify. In an institution dominated by the rights and privileges of socially conscious regiments, the newborn General Staff lacked power, influence and status.

In addition, unlike the continental armies, the British General Staff had a fairly narrow function. It did not control the army nor could a General Staff officer issue orders in his own right, unlike his German Army counterpart. Rather, the purpose of the British General Staff was to serve the commanders of a field force. In a further divergence from the continental pattern, the British deliberately split their General Staff into three sections rather than opting for a single organisation. The Chief of the Imperial General Staff controlled only the operations staff, while the Adjutant-General and the Quarter-Master General had staff organisations of their own. In effect, the British arrangement created a schism in the staff, formally dividing it into operational and administrative spheres.[72] Nor was it the British General Staff's purpose to serve as the force's thinking body.[73] In light of these differences between the practices of the British and continental General Staffs, Brian Bond and Williamson Murray correctly conclude that the British General Staff was in no way the equivalent of the German General Staff.[74]

British commentators were well aware that their version of the General Staff did not fulfil the same function as those of other armies. First published in 1889, the goal of Wilkinson's *The Brain of an Army* was to explain to Britain the purpose of the German General Staff and advocate the creation of an equivalent body for the British Army.[75] Even after its establishment, other British observers continued to call for the formation of a staff organisation based on the German model.[76] In the article 'The British Army

---

72 Anon, 'The Staff', *The Army Quarterly*, Vol. 1, No. 1, October, 1920, p. 32.

73 Dominick Graham, 'Observations on the Dialectics of British Tactics, 1904-45' in Ronald Haycock and Keith Neilson (eds), *Men, Machines & War*, Wilfrid Laurier University Press, Waterloo Ontario, 1988, pp. 60-61.

74 Brian Bond and Williamson Murray, 'The British Armed Forces, 1918-39' in Allan R Millett and Williamson Murray (eds), *Military Effectiveness, Volume II: The Interwar Period*, Allen & Unwin, London, 1988, p. 121.

75 See Wilkinson, *The Brain of an Army*.

76 Robert M Citino, *Quest for Decisive Victory: From Stalemate to Blitzkrieg in Europe, 1899-1940*, University Press of Kansas, Lawrence, 2002, p. 140.

and Modern Conceptions of War', its anonymous author displayed a well-developed understanding of the latent potential of a fully functioning, continental-style General Staff. He noted that the General Staff should be an army's thinking body. It was the only agency that had the potential to extract lessons from past actions systematically, identify the principles that governed the success or failure of commanders, adapt these observations to current conditions, and transmute these concepts into a doctrine which permeated the whole army.[77] There was no organisation within the British Army that possessed these capabilities.

The learning and leadership weaknesses of the British General Staff were evident throughout this period. Without central guidance, the advocacy and application of particular techniques remained the prerogative of the individual commander. Thus the decentralisation of tactics was often the norm: armies left corps to plan their own attacks, and corps in turn passed the responsibility to their subordinate headquarters. The result was the generation of methods that varied tremendously throughout the force. On the Western Front, whenever a division shifted to the control of a different corps, its commander and staff had to master the protocols of their new superior headquarters. Learning occurred in a decentralised, *ad hoc* manner, rather than through an established system, inhibiting efforts towards uniformity.[78]

Instead of an institutional commitment to lesson learning and force development, the British Army relied on individualism and personality-driven problem solving, within a culturally accepted framework, to guide its response to the changing nature of war.[79] Typically, a senior officer made his own decision on how a battle would be fought and imposed his ideas on his subordinates—or at least attempted to do so. The British Army did not have a doctrinal mechanism with which to provide uniformity and shape change.

---

77 Anon, 'The British Army and Modern Conceptions of War', *The Edinburgh Review*, Vol. 213, No. 436, April 1911, pp. 1184-85.

78 Tim Travers, 'Learning and Decision-Making on the Western Front, 1915-1916: The British Example', *Canadian Journal of History*, Vol. 18, No. 1, 1983, pp. 94-96.

79 Garry Johnson, 'On Winning: An Option for Change Without Decay', *The Journal of the Royal United Service Institution*, Autumn, 1991, p. 12.

The British Army's belief that doctrine was an inappropriate construct for a military organisation that had a multitude of greatly differing responsibilities across a global empire

Unlike its continental rivals, the British Army had to prepare for numerous different roles in a variety of geographic and cultural environments throughout an empire that controlled territory or had interests in every corner of the planet. For much of the nineteenth century the army occupied itself with three primary tasks: the garrison of India and other outposts of the Empire, the provision of forces for minor campaigns anywhere in the world, and home defence.[80] Increasing German belligerence saw the addition of a fourth role: the provision of an expeditionary force for a major war in Europe. With the end of the First World War, the British Army's responsibilities increased further to include the provision of troops for the territories Britain managed under the League of Nations' mandate, the occupation of part of Germany, and the intervention in Russia.

Due to the great variety in its probable tasks, senior British officers knew that the training of their forces was a far more difficult task for them than it was for the leaders of other military forces. A 1911 commentator noted that:

> the war problems of the British Empire are more complex than those of any other nation. Our interests are vaster, our responsibilities are greater, and the administration of our Army in peace, and its preparation for war are proportionally complicated.[81]

As the Second World War approached, another observer summarised the situation: 'here is of course the salient difference between us and Germany that they know what Army they will use and, broadly, how they will use it and can thus, prepare … in peace for such an event'.[82]

In the discussion that followed an address at the Royal United Service Institution in 1910, Lieutenant General Sir RSS Baden-Powell summarised the challenge the

80 *First Report of the Committee on Organization of the After-War Army*, Murray Collection, 79/48/4, Imperial War Museum, July 1919, p. 8.

81 Anon, 'The Development of the General Staff', *The Army Review*, Vol. 1, No. 1, July 1911, p. 19.

82 Quoted in Bond and Murray, 'The British Armed Forces, 1918-39' in Millett and Murray (eds), *Military Effectiveness, Volume II: The Interwar Period*, pp. 114-15.

army faced in its preparations. He highlighted the necessity of avoiding specificity in the force's regulations and instructions. Instead, he said, the army had to make its preparations on a wide basis in order to be 'prepared for almost any kind of fighting, whether it was against European Powers or against savages'. Baden-Powell concluded that 'we might be called upon to do anything'.[83] The result, as John A Nagl observes, was that 'British soldiers and officers learned on the job, and often their job had nothing to do with what they had been trained to do'.[84]

Even the issue of force modernisation resulting from the advent of mechanisation took place within the context of the army's multiplicity of missions. While describing mechanisation as imperative, Major General Ronald Adam also highlighted the need for caution. In part, this was due to mechanisation's great expense, but it was also born of concern that too great a reliance on armoured fighting vehicles would over-specialise the force. He noted that what the army really needed was a well balanced force for an average role.[85] Tanks, Adam knew, were unsuited for use in India and the lesser colonies, and the British required a heavy mechanised force only for a major war against a European opponent. With so many possible theatres of war, the tailoring of a force for a particular mission would have limited its overall utility.

In effect, Britain required two armies: a relatively light force trained and equipped for the defence of India and the Empire, and a heavy force responsible for home defence and war against a similarly armed European opponent. However, such luxury would have required a defence vote far greater than any government would have been willing to countenance. The army, therefore, had to make do as best it could, constantly balancing the conflicting requirements of the Indian and home armies, and also providing garrisons for the rest of the Empire. Torn in too many directions, the British Army's attempts to find a single focus were doomed to fail, even had any such attempts seriously been made.[86]

---

83 C W Battine, 'The Proposed Changes in Cavalry Tactics', *Journal of the Royal United Service Institution*, Vol. 54, No. 393, November 1910, p. 1433.

84 John A Nagl, *Learning to Eat Soup with a Knife: Counterinsurgency Lessons for Malaya and Vietnam*, University of Chicago Press, Chicago, 2005, p. 37.

85 *The Role of the British Army*, Liddell Hart Centre for Military Archives, Adam Collection, 2/2/1, February 1936.

86 Bond and Murray, 'The British Armed Forces, 1918-39' in Millett and Murray (eds), *Military Effectiveness, Volume II: The Interwar Period*, p. 115.

The British Army's lack of a concept of war around which to shape a doctrine

As part of the reform drive that followed the war in South Africa, British military thinkers gave considerable institutional appreciation to the need to revise the force's existing manuals in light of that conflict's lessons. The most important new publication was the 1909 *Field Service Regulations*. This volume sat at the apex of the British Army's manual hierarchy, and all lesser writings were to conform to its dictates.

Given the stature of the *Field Service Regulations*, it would be tempting to borrow from contemporary language and describe this as the British Army's first capstone manual of doctrine. However, such a declaration would be a mistake since the manual does not articulate a concept of war. Had the army possessed a defined style of waging war then, according to the *Field Service Regulations*, it must have been the pursuit of the decisive battle. This was hardly a unique objective for an Edwardian-era army. Yet the manual's sections on the conduct of battle are presented so poorly and vaguely as to be virtually useless. How a commander and his subordinates were to achieve victory through a decisive battle was left largely to their own interpretation, although the manual expected them to base their actions on unidentified and undefined 'principles of war'. The *Field Service Regulations* also failed to solve the key tactical problem of the era: how infantry were to assault across the fire-swept killing zone that lay in front of the enemy's position.[87]

Instead of clarity, the *Field Service Regulations* was filled with obfuscation. Rather than sound instructions, commanders received aphorisms. Unable to proffer reasoned tactics, the *Field Service Regulations* told its readers that 'skill, better organization, and training, and above all a firmer determination in all ranks to conquer at any cost, are the chief factors of success' and that 'decisive success in battle can only be gained by vigorous offensive'.[88] At the first conference of the General Staff that followed the publication of the *Field Service Regulations*, the problem of infantry assault tactics

87 Dominick Graham, 'Sans Doctrine: British Army Tactics in the First World War' in Timothy Travers and Christon Archer (eds), *Men at War: Politics, Technology and Innovation in the Twentieth Century*, Precedent, Chicago, 1982, p. 73; and Citino, *Quest for Decisive Victory*, p. 54.

88 War Office, General Staff, *Field Service Regulations (1909)*, p. 107.

was placed firmly on the agenda.[89] While the conference failed to resolve the issue, it did produce further aphorisms such as 'to conquer is to advance'.[90]

Commenting on the British Army's manuals, Shelford Bidwell and Dominick Graham observe that the *Field Service Regulations* and the subordinate arms manuals provided no assistance in explaining how the army should fight. Instead they were written in general language and avoided concrete examples.[91] One damning description described the army's manuals as being as useful to modern soldiers 'as cuneiform inscriptions on a Babylonian brick'.[92] This was the basis of General Hippolyte Langlois' oft-repeated description of British manuals as flawed providers of doctrine. He wrote that, while the British Army had produced excellent manuals, which he admitted were as good as those of other European armies, they lacked a common doctrine. He continued, 'without a doctrine, text books are of little avail: better a doctrine without text books than text books without a doctrine'.[93]

Admittedly, Langlois, whose book was first published in 1904, was commenting on the state of the British Army's doctrine on the eve of the war in South Africa. However, his words resonated throughout the Edwardian Period and were repeated by a number of authors in their own writings on the intellectual development of the army. Langlois' observations thus remained pertinent even after the publication of the *Field Service Regulations*.[94]

When the army's manuals did attempt to provide guidelines for action, these took the form of abstractions, or what Graham has referred to as 'the usual clichés'. He adds that 'divisional commanders were at liberty to interpret the clichés of the manuals as

89 *Report of a Conference of General Staff Officers at the Staff College, 17th to 20th January 1910*, HMSO, London, 1910, pp. 25-28.

90 Ibid., p. 28.

91 Bidwell and Graham, *Firepower*, p. 19.

92 Quoted in D S Graham, 'The Ascendancy of Firepower' in J A English, J Addicott and P J Kramers (eds), *The Mechanized Battlefield: A Tactical Analysis,* Pergamon-Brassey's, Washington, 1985, p. 56.

93 H Langlois, *Lessons from Two Recent Wars: The Russo-Turkish and South African Wars*, HMSO, London, 1909, p. 145.

94 For examples of Langlois' appeal to British authors see Anon, 'The General Staff in India', *The Army Review*, Vol. 2, No. 1, January 1912, p. 21; and Fuller, *Training Soldiers for War*, p. 45.

they wished, and few agreed'.[95] It probably did not help matters that some authors borrowed liberally from foreign military publications, which they translated 'with artistic merit into British use'.[96] Rather than 'capstone', the term generally applied to an armed forces overarching doctrinal manual, the most apt description of the style and contents of the British *Field Service Regulations* is 'platitudinous'.

Reinforcing the lack of enthusiasm with which newly minted manuals were received was an institutional bias that denigrated book-learning. British officers of the First World War era did not enjoy a reputation for professional reading. The historian Tim Travers retells the story of Sir John French, the future Commander-in-Chief of the British Expeditionary Force, who, while at the Staff College, having found Edward Hamley's *Operations of War* impenetrable, never again asked the War Office librarian for another book.[97] At the opposite end of the spectrum, J F C Fuller's reading habit was part of the reason his fellow officers considered him peculiar. In return, it was during his first posting that Fuller developed a career-long contempt for the intellectual deficiencies of his peers. He read philosophy while his brother officers passed their time either hunting and fishing, or talking about hunting and fishing.[98] Fuller remembers that one of the rituals of his regiment's annual inspection was an officer's presentation of his collection of manuals and regulations. Each officer produced these volumes tied with string, to be untied and checked to ascertain that they had received a proper education. Each officer then rebound his stack of books and put them away until the next year.[99]

Not only did the army view intellectualism with suspicion, but some senior British officers were actually hostile to any suggestion that they had derived professional knowledge from a book.[100] Instead they insisted that actual, not theoretical, field experience was the only legitimate means to acquire the necessary skills of warfighting

---

95 Graham, 'Sans Doctrine', p. 75.

96 Ibid., pp. 19, 38-39.

97 Tim Travers, *The Killing Ground: The British Army, The Western Front and the Emergence of Modern Warfare, 1900-1918*, Unwin Hyman, London, 1987, p. 40.

98 J F C Fuller, *Memoirs of an Unconventional Soldier*, Ivor Nicholson and Watson, London, 1936, p. 6.

99 Brian Holden Reid, 'Introduction: Is There a British Military Doctrine' in J J G Mackenzie and Brian Holden Reid (eds), *Central Region vs Out-of-Area: Future Commitments*, Tri Service Press, Camberley, 1990, p. 6.

100 Travers discusses the Army's anti-intellectualism in *The Killing Ground*, pp. 38-40.

and leadership. Philip Christison's first assignment after Staff College found him at the Military Training Branch at the War Office. His chief was General Sir Harry Knox, one of the army's most conservative senior officers. Knox, Christison recalled, greeted his arrival with the exclamation, 'I hope you are not one of these book-writing chaps; I have suffered from them'. Christison wisely replied in the negative, to which Knox exhorted 'Thank Heaven!'[101]

The failure of British manuals to instil a common vision of the army, overcome tribalism, and inculcate a spirit of cooperation between the separate arms provides compelling evidence of the error in accepting the existence of manuals as proof of a military organisation's indoctrination. Manual publication is easy, but the path to doctrinal inculcation is difficult and fraught with institutional obstacles. Brian Holden Reid observes that 'publishing a doctrinal pamphlet or circulating a paper is no more proof of the acceptance of a doctrinal policy than shouting its conclusions from the roof of the old War Office'.[102]

In the final analysis, doctrine is a matter of faith, and for much of the British Army's modern history it did not have sufficient disciples committed to its texts to rely upon manuals with which to support a doctrine. Timothy Lupfer's conclusion on this point is extremely apt. He writes: 'doctrine that influences nothing beyond the printing press is stillborn'.[103]

The British Army's fears that doctrine would lead to stereotyped thinking and degrade the role of commanders in battle

While the three factors described above provide ample evidence of the serious structural barriers to doctrinal innovation that existed within the British Army, there remained one further and even more significant impediment. Any attempt at indoctrination also had to overcome powerful cultural factors which were integral to the very spirit of the institution. These took the form of an army-wide fear of stereotyped thinking and a desire to preserve the importance of the commander in battle. In opposition to indoctrination, the officer corps consistently advocated a preference for

---

101 *Memoirs*, Christison Collection, ID No. 82/15/11, Imperial War Museum, p. 87.

102 Brian Holden Reid, 'War Fighting Doctrine and the British Army' in Brian Holden Reid, *A Doctrinal Perspective, 1988-98*, Strategic & Combat Studies Institute, Camberley, 1998, p. 14.

103 Timothy T Lupfer, *The Dynamics of Doctrine: The Changes in German Tactical Doctrine During the First World War*, Combat Studies Institute, Fort Leavenworth, 1981, p. viii.

independent thinking and avoided anything that suggested a stereotyped response to situations that commanders might face. The shorthand expression for this attitude was the advocacy of the use of 'common sense'. Throughout the period of the First World War and even after, efforts towards greater standardisation of thought or action were invariably countered by cries for the development of independent thinking, 'common sense', and the preservation of a commander's prerogative to train his unit in a manner that he alone saw fit.

Given the army's strong adherence to these values, the arrival of the *Field Service Regulations* in 1909 and the subsequent publication of a variety of arms manuals made little headway in overcoming the force's reluctance to accept doctrine. The British officer's antipathy towards imposed ideas was so strong that the army inserted a phrase at the beginning of many of its publications enshrining the right of a commander not to follow the recommendations outlined in the manual. When issued in 1923, for example, *Notes on Elementary Tactical Training* contained the statement 'it would be disastrous to all good training should officers imagine that a solution suggested from the War Office or elsewhere can be made applicable to every situation'.[104]

At the 1911 annual conference of the General Staff, the first item on the agenda was the army's need for a doctrine. Doctrine's advocate was a relatively junior officer, Captain CAL Yate, then serving in the War Office on the General Staff. He called for the publication of what he referred to as a 'Manual of Applied Tactics'. While he did not employ the term 'doctrine', the manual he proposed would supposedly have fulfilled that function in combination with the *Field Service Regulations*. Yate gave as his rationale the fact that the army did not presently have any publication which dealt with tactics in practical terms. Instead, officers had to resort to 'crammers' or books published by foreigners.[105]

The conference allowed Yate's suggestion a fair hearing before rejecting his idea. The participants admitted that none of the army's existing manuals served the proposed purpose, and they suspected that there were benefits to be gained from its publication. However, they also voiced serious concerns, and these ultimately proved Yate's undoing. Major General ES May voiced the objection that, with such widespread imperial commitments, the British Army must prepare for a great variety of

---

104 War Office, *Notes on Elementary Tactical Training*, HMSO, London, 1923, p. 2.

105 See *Report of a Conference of General Staff Officers at the Staff College, 9th to 12th January, 1911*, pp. 5-9, 28.

circumstances and unforseen emergencies. He concluded that officers who relied on a manual on tactics ran the risk of being seduced by stereotypical thinking, a problem which the institution, May observed, 'must guard against'.[106] Brigadier Generals F J Davies and R C B Haking also spoke against the notion. They observed that it would be difficult to produce a book that could meet every tactical situation and, even if the army managed to do so, keeping it current would be impossible. The consensus of the conference's participants was to refuse to allow the proposed manual official status, although they did not object to individual officers writing their own works on the subject.[107]

Outside General Staff channels, an even broader debate was taking place in defence journals and commercial publications. For example, in *Training Soldiers for War*, Fuller declared that the *Field Service Regulations* represented the army's doctrine. However, he immediately undercut his claim by summarising this alleged doctrine as 'we will seek out the enemy and destroy him'. This was hardly useful, and would be of little assistance to his readers. It was also as much a platitude as those that filled the official manuals.[108]

Ultimately, two articles advocating the adoption of doctrine provoked the most heated exchange of views on the matter. The articles appeared anonymously in *The Edinburgh Review* in 1911 and 1912, clearly from the same pen, and respectively titled: 'The British Army and Modern Conceptions of War' and 'The Place of Doctrine in War'.[109] Commenting on Britain's failings during the South African War, the author concluded that 'a sound, comprehensive, all-pervading doctrine of war is as important to an army as its organization, that it is the soul without which an army is but so much inert matter and that in this vital respect the British Army of 1899 … was completely lacking'.[110] In both articles the author drove home the point that the British Army needed a doctrine in the continental style. He believed that the speed and scale of

106 Ibid., p. 7.

107 Ibid., p. 28.

108 See Fuller, *Training Soldiers for War*, p. 45.

109 See Anon, 'The British Army and Modern Conceptions of War', pp. 321-46 (the article was subsequently reprinted in the *Journal of the Royal United Service Institution*, Vol. 55, September 1911); and Anon, 'The Place of Doctrine in War', *The Edinburgh Review*, Vol. 215, No. 439, January 1912, pp. 1-30.

110 Anon, 'The British Army and Modern Conceptions of War', p. 324.

modern war required armies to have greater unity of action. In particular, he considered that the British required standardisation to the point where a subordinate would know how to respond to a situation without recourse to his commanding officer. Moreover, he continued, commanders needed to be confident that their subordinates would respond in an expected manner. These were words which echoed Moltke's beliefs.

A number of officers responded to these essays. The most powerful counter-argument came from the pen of Brigadier Thomas Capper.[111] Capper was incensed by the article's proposals. He labelled the doctrine advanced in the *Edinburgh Review* as one by which 'we must fight on a certain given system, if we fight at all'. Capper found it 'extraordinary' and 'inconceivable' that a commander could wage war using only one method or version. He believed that the attraction of doctrine was that it provided people with hard and fast rules. In his opinion, however, this was its very drawback. Capper pointed out that Britain, unlike Germany and France, had a multitude of different military tasks to prepare for, and could not afford a doctrine devoted to a single method. Capper also asserted that the proposed doctrine would militate against thinking, and would encourage categorical solutions to training problems. Instead of promoting the force's intellectual development, doctrine would have the opposite effect of stifling independent thought.

In addition to the belief that doctrine would lead to stereotyped thinking, its consideration prompted the voicing of other visceral fears by British Army thinkers. A recurring concern was that doctrine limited the options of commanders in waging war and, as a consequence, would lead to the death of generalship. One officer wrote that 'the army of the future is to be reconciled to the loss of a leader by the acquisition of a "doctrine"'.[112] Another objection held that doctrine was a current fashion which would undercut the centrality of the principles of war in the conduct of operations. The army accepted that the principles of war were timeless, but saw doctrine as a set of manufactured and imposed rules. The problem with rules was their tendency to induce rigidity and slavish adherence and thus they would become transmuted

---

111 For Capper's response, see 'Response by Capper to Article in *The Edinburgh Review*', Capper Collection, 2/4/20, Liddell Hart Centre for Military Archives.

112 'Veto', 'Is "Doctrine" a Delusion?', *United Service Magazine*, Vol. 66, November 1912, p. 178.

into dogma. Instead of doctrine, what the army needed was an understanding of the principles of war and the experience to implement them.[113]

The outbreak of the First World War put an end to the discussion on doctrine that had followed the publication of the *Field Service Regulations*. Fuller himself highlighted the transient nature of the army's desire for indoctrination. Shortly after the conflict's commencement, in an essay titled 'The Tactics of Penetration: A Counterblast to German Numerical Superiority', Fuller contradicted the favourable opinions on doctrine which he had so recently espoused in *Training Soldiers for War*. The mercurial Fuller now wrote, 'I have no doctrine to preach, for I believe in none'. He continued, 'A physician who is slave to a doctrine ... ends by killing his patients ... If there is a doctrine at all then it is common sense, that is, action adapted to circumstances.' If Fuller had a doctrinal philosophy it was now: 'Know your weapons. Understand your enemy.'[114]

Discussion over whether or not to seek greater uniformity of thought and procedure through the inculcation of a doctrine or to support the existing policy of independent thinking, was not confined to British shores. In 1916 the Indian Army campaigned against the Mohmand people in the North-West Frontier territory of what is now Pakistan. The operational styles of the two divisions involved, the 1st (Peshawar) and 2nd (Rawalpindi), were so different that they became the subject of a conference held in early 1917. One of the conclusions reached was that the army lacked uniformity in the way it interpreted the principles outlined in the mountain warfare section of the *Field Service Regulations*. Major General William Bunbury argued for more definite rules.[115] He noted that:

> every teacher of mountain warfare and every writer on the subject appears to have his own ideas, not so much as to principles but as to how they should be

---

113 'Veto', 'Is "Doctrine" a Delusion?', p. 179; 'Mousquetaire', 'Heretical Doctrines of War', *United Service Magazine*, Vol. 66, January 1913, p. 397; and AWA Pollack, 'Strategy in Theory and Practice', *United Service Magazine*, Vol. 67, July 1913, p. 421.

114 J F C Fuller, 'The Tactics of Penetration: A Counterblast to German Numerical Superiority', *Journal of the Royal United Service Institution*, Vol. 59, November 1914, p. 389; Fuller, *Training Soldiers for War*, p. 45.

115 T R Moreman, *The Army in India and the Development of Frontier Warfare, 1849-1947*, Palgrave, London, 1998, pp. 99-100.

> applied, and the result is that one finds different teaching on certain points in every brigade and even in the battalions of a brigade.[116]

However, the commander of the Peshawar Division, Major General Sir Frederick Campbell, warned of the danger of precise rules. Other officers opposed additional manuals and instead spoke of the need to base operations on the principles of war. The conference produced a compromise which advocated reliance on standing orders as a means to impose some degree of uniformity. However, the issue of standing orders was the prerogative of every commander, thus assuring that every operation, formation and unit produced its own, different, standing orders.[117]

The passage of time did little to dispel the fear among senior officers that doctrine would invariably lead to stereotyped thinking. Writing in his history of the 8th Army in North Africa, published in 1963, Lieutenant General Sir Francis Tuker dismissed the value of doctrine in language similar to that of his predecessors at the General Staff Conference of 1911. Rather than advocating doctrine, he argued that the British Army should continue to avoid it 'like the plague' because, he concluded, 'it will bind men's minds'.[118] Instead of classwork or manual reading, Tuker expected officers to learn through experience in the field and by commanding real units.

---

116 Ibid.

117 Ibid.

118 Francis Tuker, *Approach to Battle: A Commentary, Eighth Army – November 1941 to May 1943*, Cassell, London, 1963, pp. 389-90.

Mastering the Western Front

The military situation on the Western Front during the First World War is commonly represented as a stalemate. The combatants—Germany on one side and the combined forces of Britain and its dominion and colonial troops, France, Belgium, and later the US on the other—dug extensive trench systems from which their troops controlled the intervening no man's land with deadly fire. The result was a continuous defensive network that ran from the English Channel to neutral Switzerland, and front lines that barely moved for much of the war, despite the enormous expenditure of materiel and men.[119]

The main challenge facing the attacker came from the high concentration of manpower in a relatively small area, the strength of the defences, and the deadly effects of modern weapons. Each side could only restore mobility and break the stalemate if it learned how to negate its opponent's defensive firepower. This would allow the attacking infantry to cross no man's land in sufficient numbers to breach the defensive works and break through the opposing lines. How this was to be accomplished was the crucial tactical question of the war, and one that took almost four years to answer.

Of the First World War's major protagonists, Britain was the least prepared for the nature and intensity of combat on the Western Front. Traditionally, the British Army's primary strategic requirement was to garrison the Empire and, while its equipment was modern and its troops well trained, it was a small organisation both relative to Britain's population base and to its much larger continental rivals. From the British perspective, the dominant theme of the years 1914 and 1915 was the rapid expansion of the army. Equally important was a vast increase in the nation's armament production. In both tasks Britain succeeded, and in doing so created an army that was capable of waging war on a continental scale against a European opponent. Just one example that highlights the extent of the British Army's growth is the increase in its establishment of heavy artillery. In 1914 the British Expeditionary Force in France could call upon just

119 Jeremy Black, *A Military History of Britain: From 1775 to the Present*, Praeger Security International, Westport, 2006, p. 113.

six batteries of heavy guns for support. By contrast, four years later at the war's end, it fielded 440 batteries—a 73-fold increase.[120]

Force expansion, however, was only one part of the equation the British needed to solve in order to achieve victory. After the institution of the trench system, attempts to defeat the Germans first relied on the simple application of mass with which to obliterate the enemy's soldiers and defensive works. Britain's offensives throughout 1915 were relatively minor affairs, largely due to inadequate numbers of men and materiel. The real test of this methodology came with the Battle of the Somme, which served only to reveal its inadequacy. On 1 July 1916, the offensive's first day, the British lost almost 20 000 dead and 40 000 wounded, despite a lengthy preliminary bombardment. The British Army's attempt to break through the German trench-lines by force was a costly failure.

While the Somme was a catastrophe that seared the collective memory of the nation, it also revealed numerous deficiencies in the British technique of warfighting. In a sense, it was a turning point. Over the next two years the British introduced a series of technological and tactical innovations that would radically alter the army's methods and make possible the victories of the second half of 1918 that forced the Germans to sue for peace. In the process, the British also discovered the fundamentals of modern combined arms warfare.

The tactical solution to the impasse of the Western Front emerged in the more sophisticated application of offensive firepower with which to control the battlefield, and thereby allow the infantry to cross no man's land and close with the enemy. One of the key lessons of the Battle of the Somme was the need to locate and silence the enemy's guns. Even before the battle's end the British had made great strides in the development of an effective counter-battery system whose centrepiece was to be known as the Counter-Battery Staff Office. Into this corps-level operations centre flowed intelligence gathered by the sound-ranging, flash-spotting and aerial reconnaissance organisations. The Counter-Battery Staff Office then plotted the location of the enemy's batteries and developed fire plans with which to silence

120 Ibid., p. 113.

specific guns as required. A serving gunner of today would recognise the principles of counter-battery fire devised initially by the British in the First World War.[121]

The British also took the lead in the development of the tank. This novel weapon made its first appearance in the September phase of the Somme Offensive. Rushed to the front in too few numbers and deployed on unsuitable terrain, the tank's debut proved far from impressive. As a weapons system, however, it showed sufficient promise to prompt the British Commander in France, General Douglas Haig, to call for the production of 1000 vehicles. Henceforth, the British incorporated armoured vehicles into the plans for most of their offensives.[122]

Another area in which the British excelled was chemical warfare. Gas had been a German innovation, used for the first time in significant quantities at the Battle of 2nd Ypres in April 1915. Germany's chemical industry was Europe's largest and, consequently, it enjoyed an early advantage in gas warfare. Britain responded, however, with a rapid expansion of its own chemical industry, including the introduction of a number of chemical agents and delivery techniques. The British Livens Projector was to become the conflict's most feared gas delivery system. By the second half of 1918 Britain, along with its allies France and the US, had won the chemical war and it was the German front-line soldier who bore the brunt of its effects.[123]

While expansion of munitions production and improvements in weaponry were vital, of even greater importance to Britain's victory over Germany were advances in its small unit fire and movement tactics. In the earlier stages of the war, the British battalions had attempted to march across no man's land in linear formations similar to those of the Napoleonic Era. By 1918, they had learned to manoeuvre forward in

---

121 For an account of advances in British gunnery techniques, see Albert Palazzo, 'The British Army's Counter-Battery Staff Office and Control of the Enemy in World War I', *The Journal of Military History*, Vol. 63, No. 1, January 1999, pp. 55-74.

122 Stephen Peter Rosen, *Winning the Next War: Innovation and the Modern Military*, Cornell University Press, Ithaca, 1991, p. 122. A number of works cover the British development of the tank. Among the best are David J Childs, *A Peripheral Weapon? The Production and Employment of British Tanks in the First World War*, Greenwood Press, Westport, 1999; and J P Harris, *Men, Ideas and Tanks: British Military Thought and Armoured Forces, 1903-1939*, Manchester University Press, Manchester, 1995.

123 For the British story on the development of gas, see Albert Palazzo, *Seeking Victory on the Western Front: The British Army and Chemical Warfare in World War I*, University of Nebraska Press, Lincoln, 2000.

mutually supporting sections, each platoon advancing in fire teams supported by its organic light machine-guns and mortars.

The Australian Corps under Lieutenant General John Monash played a critical role in the development of British combined arms tactics. The Australian-planned and executed Battle of Hamel was a small-scale exposé of the new techniques that the British Army as a whole had developed and incorporated into its method of waging war. The battle was a textbook success and became a model for future British operations.[124]

The first major test of the evolved British tactics was the Battle of Amiens. The battle's commencement was heralded by the firing of a counter-battery program that silenced almost all of the German Army's guns on the attack front. British and French tanks then rolled forward, crushed the belts of barbed wire that protected German positions, and eliminated enemy machine-guns and strongpoints with direct fire. The infantry followed in the tanks' wake, advancing in bounds with sections mutually supporting one another with fire and movement. The British ruptured the German front and penetrated deep into the enemy's lines. By the end of the first day the British and the supporting French troops had taken over 30 000 prisoners and captured 500 guns.[125] The German General Erich von Ludendorff subsequently called 8 August 1918 'the black day of the German Army'.[126]

The Germans never recovered from the shock of Amiens. Instead, the battle marked the commencement of a series of offensives collectively known as the Hundred Days' Campaign. During this period the armies of Britain, France and the US allowed their opponent no respite. When the British IV Army ruptured the Beaurevoir Line in early October, the Germans lost their last fortified line in France. The German Government's response was to seek talks that would result in the declaration of the Armistice.

---

124 J P Harris with Niall Barr, *Amiens to the Armistice: The BEF in the Hundred Days' Campaign, 8 August-11 November 1918*, Brassey's, London, 1998, pp. 56-57.

125 Ibid., p. 116.

126 Quoted in S F Wise, 'The Black Day of the German Army: Australian and Canadians at Amiens, August 1918' in Peter Dennis and Jeffrey Grey (eds), *Defining Victory, 1918*, Army History Unit, Canberra, 1999, p. 3.

Space constraints in this study paper allow only the briefest summary of the British Army's transformation[127] and little mention of the innovations of the other combatants which were also considerable. The crucial point is that the British Army that defeated the Germans in 1918 bore little resemblance to the force that entered the conflict in 1914. The British Army transformed itself from a force designed primarily for the defence of the Empire into one capable of resisting and defeating a major European opponent. British commanders and political leaders recognised the need to change and implemented the necessary modifications. Throughout this transformation the British examined the challenges and problems of combat on the Western Front, sought solutions, and incorporated new technologies and improved tactics into their method of waging war. Most significantly, they did this without recourse to doctrine.

127 There is now a considerable body of literature on British Army innovation in the First World War. A brief overview can be found in G D Sheffield, 'The Indispensable Factor: The Performance of British Troops in 1918' in Dennis and Grey, *Defining Victory, 1918*, pp. 72-95. Longer works include: Paddy Griffith, *Battle Tactics of the Western Front: The British Army's Art of Attack, 1916-1918*, Yale University Press, New Haven, 2004; Tim Travers, *How the War Was Won: Command and Technology in the British Army on the Western Front 1917-1918*, Routledge, London, 1992; Childs, *A Peripheral Weapon?*; and Palazzo, *Seeking Victory on the Western Front*. For the German perspective of innovation on the Western Front, see Lupfer, *The Dynamics of Doctrine*; and Bruce I Gudmundsson, *Stormtroop Tactics: Innovation in the German Army, 1914-1918*, Praeger, New York, 1989.

The British Army's alternative to doctrine—institutional ethos

The analysis presented in this study paper will create a conundrum for contemporary military professionals and thinkers. The British Army underwent a massive transformation and improvement in capability over the course of the First World War. However, it did so in the complete absence of a doctrine, either before, during or after the conflict. Perhaps the final statement regarding its absence should belong to the Chief of the Imperial General Staff in 1939, Field Marshal Sir Edmund Ironside. As war with Germany loomed again, he identified as one of his most pressing tasks the development and inculcation of a doctrine. This, he commented, had been absent since the end of the last war.[128] The British Army had waged the First World War without a doctrine and, with the passage of twenty years and the next great conflict looming, its commander declared that it still did not have one.

If doctrine, as most contemporary military professionals and thinkers insist, is the essential intellectual framework through which an army interprets the nature of war and makes innovations in its methodology, then the British should not have been successful in their efforts to implement changes during the First World War. The flaw in the logic of the advocates of doctrine is the assumption that it is the sole mechanism with which a military organisation can interpret the nature of war, and through which it can disseminate and inculcate change. A more correct conclusion in the case of the First World War-era British Army is to suggest that other mechanisms of change are possible.

In fact, the British Army did possess a mechanism for modifying its method of waging war—a non-doctrinal mechanism—that provided the force with a unifying effect and the means to identify problems, pose solutions and implement change. This mechanism was an institutional 'ethos'. There can be little doubt that the advancement of ethos as an alternative to doctrine will pose a problem for many military practitioners

128 'Ironside to Lindsay', Imperial War Museum, Ironside Collection, I.D. No. 92/40/1, 22 June 1939.

and thinkers. Doctrine-based analysis has its own vested interests, both within military organisations and amongst those who study them. However, claims that assert the existence of a British doctrine of war, or herald the publication of the *Field Service Regulations* and other manuals as the arrival of doctrine, are not supported by the evidence. During the era of the two world wars there were numerous instances of British military leaders either rejecting the need for doctrine or calling for its creation, and strong evidence that the institution and its senior thinkers did not consider that it yet possessed one. These instances have been ignored by many scholars in their determination to identify the force's doctrine and fit the British Army's intellectual mechanisms into contemporary military theory.

Observers continue to disagree, however, on whether it is possible for a culturally based institutional ethos to exist. While commentary on this subject is limited, there is some evidence in favour of the idea. On the eve of the war in South Africa, Fuller concluded not only that the force did not possess a doctrine, but also that it lacked a unifying spirit.[129] More recently, in another refutation of the existence of a British Army doctrine, Brian Holden Reid observes that it is 'crystal clear ... [that] the British Army does not embrace a "philosophy" ... which animates the actions of all soldiers'.[130] More promisingly, David French deduces that the British Army did not need a doctrine because it relied on something else for its unity. He describes this as the 'common character' of the army's members.[131]

French's observation resonates with the ideas of several of the British Army's senior officers. Shortly after the end of the First World War, for example, Lieutenant General Ivor Maxse proposed that instead of books, the army should rely on its training to enshrine the lessons of the recently ended conflict in a national habit which he expected would become known as a 'British war tradition'. He believed that if the army attempted to codify this tradition in published form, it would die out.[132] Maxse's comment is interesting because he was an opponent neither of learning nor of the

129 Fuller, *Training Soldiers for War*, p. 43.

130 Holden Reid, 'Introduction' in Mackenzie and Holden Reid (eds), *Central Region vs Out-of-Area: Future Commitments*, p. 1.

131 David French, *Raising Churchill's Army: The British Army and the War against Germany 1919-1945*, Oxford University Press, Oxford, 2000, p. 50.

132 'Notes on Organisation of Future British Army', Imperial War Museum, Maxse Collection, ID No. 69/53/17, File 66/1, 1919.

written word. After all, it was Maxse who oversaw B H Liddell Hart's post-war revision of the *Infantry Training* manual. Yet Maxse also recognised that the army could make use of an instrument that was more powerful than the printing press and thereby inculcate a distinctive British spirit of warfighting throughout the force.

One of Thomas Capper's reasons for rejecting doctrine during the Edwardian-era debate on its utility was that he considered it, in the British case, not only unnecessary but also potentially dangerous. Instead of a European-style doctrine, Capper advocated a spiritual doctrine based on the 'highest type of moral and physical courage'. It was this spirit, he wrote, which infused 'into our very blood, a high spirit of war'. Capper defined this spirit as 'the true warrior spirit'. Employing language similar to Maxse's, Capper argued that what the army required 'is not a doctrine at all, it is instead the very spirit of war which gives life to an army, without which an army is no army at all'.[133] Capper added:

> The history of the world reveals this cycle—the birth of a spirit; the promulgation of that spirit, the crystallizing of that spirit into doctrine; the glorification of the doctrine at the expense of the spirit; eventually the destruction of the spirit in the life of doctrine.[134]

The risk to the army of a continental style of doctrine, Capper suggested, was that 'doctrines soon produce doctrinaires, and doctrinaires soon produce dogma'.[135] At the 1910 conference of the General Staff, Lieutenant Colonel W T Furse also spoke of the need to preserve a military spirit. He feared that the army would 'destroy this spirit if you put into our regulations anything of this sort about the recognised limit to what can be expected'. Furse's emphasis lay on an unwritten spirit rather than regulations.[136]

---

133 *Response by Capper to article in* The Edinburgh Review, Liddell Hart Centre for Military Archives, Capper Collection, 2/4/20.

134 Ibid.

135 Thomas Capper, *The Doctrine of a 'Doctrine'*, Liddell Hart Centre for Military Archives, Shea Collection, 2/5.

136 *Report of a Conference of General Staff Officers at the Staff College 17th to 20th January, 1910*, p. 15.

In a manner reminiscent of the British Constitution, the army's spirit (or ethos) existed, albeit unwritten. Admittedly, this complicates its definition; yet the absence of documentary confirmation denies neither the existence of ethos nor its importance. It does mean however, that as with a solar eclipse, investigators must be willing to employ indirect means of observation.

All military organisations, like their parent societies, possess distinctive institutional cultures. Culture played an especially critical role in defining the nature and potential of the British Army. The cultural values of the army during the era of the two world wars were also the wellspring from which it derived its institutional ethos. Consequently, unlike doctrine, which is an intellectual construct that a military organisation deliberately fabricates to meet a particular need, ethos is a cultural construct that is intimately woven into the fabric of a military institution's spirit, history and tradition, and which infuses its members with a shared understanding, conception and perspective of war.

Since ethos is representative of an organisation's cultural values, it is these values that give rise to the unifying factors that shape a military force into an institutional whole. A number of scholars have identified the characteristics that distinguish the British Army from the military organisations of other nations and which, thereby, have shaped its unique ethos. One frequent observation is that the British Army is 'a pragmatic, empirical institution'.[137] Brian Holden Reid comments that the army:

> traditionally depreciated any notion that war can be regarded as the manipulation of fixed quantities, it has always stressed the value of pragmatic—or vicarious—experience, in which the human element is a critical component.[138]

---

137 J M Bourne, 'Haig and the Historians' in Brian Bond and Nigel Cave (eds), *Haig: A Reappraisal 70 Years On*, Leo Cooper, Barnsley, 1999, pp. 7-8.

138 Brian Holden Reid, *War Studies at the Staff College, 1890-1930*, Strategic & Combat Studies Institute, Camberley, 1992, p. 2.

Stephen Hart notes that in the Second World War, 'following a long tradition', the British 'relied heavily on extemporaneous pragmatism rather than on a formally expressed doctrine of war'.[139]

The British Army's widespread reliance on empiricism took the shape of a preference for improvisation in lieu of theory-based analysis.[140] Scholars and military professionals have condemned empiricism as simply an excuse for 'muddling through', a term of disdain in use as early as 1913.[141] Yet, such a free-thinking approach did offer some benefits. For example, during the First World War, the army's willingness to experiment widely without pausing to develop theoretical underpinnings enabled the British to quickly deploy new weapons systems, including the tank, and to counter the enemy's early advantages as it did in the conduct of chemical warfare.[142]

The army's leaders also possessed firm ideas on an institutional code of behaviour, both for themselves and for their men. Having surveyed a number of service journal articles, Travers concludes that the British officer valued 'experience, surprise, initiative, imagination, genius, intelligence, common sense, and so on'.[143] Major General Sir Hugh Sandham expected his men to possess 'courage, devotion to duty, determination and endurance'.[144] General Sir Burnett-Stuart described the army as possessing 'amazing loyalty, courage and endurance, disciplined on the sure basis of mutual confidence and respect between officers and men'. He continued that the army remained 'an institution steadfast, masculine, and essentially British'.[145]

The British insisted that it was their character traits that set them apart from the armies of other nations. In discussing the army's post-war infantry organisation, Maxse expressed his belief that German and French establishments did not suit the 'British temperament'. He continued, 'we have in fact no need to copy either the French

---

139 Stephen Ashley Hart, *Montgomery and 'Colossal Cracks': The 21st Army Group in Northwest Europe, 1944-45*, Praeger, Westport, 2000, p. 79.

140 Holden Reid, 'War Fighting Doctrine and the British Army', p. 12.

141 H Musgrave, 'Various Aspects of Preparation for War', *The Army Review*, Vol. 5, July 1913, p. 47.

142 Childs, *A Peripheral Weapon?*, p. 163. See also Palazzo, *Seeking Victory on the Western Front*.

143 Travers, *The Killing Ground*, p. 41.

144 J O Coop, *The Story of the 55th (West Lancashire Division)*, Daily Post, Liverpool, 1919, pp. 8-9.

145 Burnett-Stuart, *Memoirs*, Liddell Hart Centre for Military Archives, Burnett-Stuart Collection, 6/20.

or the Germans in this particular matter'.[146] At the 1910 conference of the General Staff, several officers spoke out against the adoption of German ideas. These officers regarded British methods as best suiting the British Army.[147] This was not just the work of cultural bias. It was the officers' recognition that German methods were different and unsuitable for the British temperament. In a post-First World War assessment, Lieutenant Colonel W H F Weber argued that national character was the ultimate arbiter of victory.[148]

Reinforcing the unifying effect of these shared traits was the homogeneity of the officer corps' origins. The British Army drew its leaders from a narrow, definable, and recognisable stratum of the nation's manhood. This resulted in an officer corps that contained a disproportionate number of members from the aristocracy and gentry.[149] Byron Farwell observes that, on arrival at Sandhurst or Woolwich, a new officer cadet entered an environment composed of other young men whose backgrounds were identical to his own. Instructors and cadets spoke the same language and possessed the same vocabularies, manners, attitudes and beliefs.[150] Insufficient salaries and the cost of an officer's maintaining himself to regimental standards created a monetary barrier to wider admission. As a result, the narrow pool from which the British Army drew its military leaders encouraged a similarity of outlook and station.[151]

Even during the British Army's massive expansions through the world wars, its leaders strove to maintain the social composition of its officer corps. As the demand for officers necessitated the extension of recruitment to include non-traditional classes,

---

146 Maxse to Wilson, Imperial War Museum, Maxse Collection, ID No. 69/53/17, File 66/1, 3 January 1919.

147 *Report on a Conference of General Staff Officers at the Staff College 17th to 29th January 1910*, pp. 10-14.

148 W H F Weber, *A Field Artillery Group in Battle: A Tactical Study Based on the Actions of 2nd Brigade, R F A during the German Offensive, 1918, the 100 Day's Battle, and the Battle of Cambrai, 1917*, Royal Artillery Institute, London, 1919, p. x.

149 M A Ramsay, *Command and Cohesion: The Citizen Soldier and Minor Tactics in the British Army, 1870-1918*, Praeger, 2002, p. 58.

150 Byron Farwell, *Mr Kipling's Army: All the Queen's Men*, Norton and Company, New York, 1981, p. 145.

151 For a study of the social composition of the officer corps see Keith Simpson, 'The Officers' in Ian F W Beckett and Keith Simpson, *A Nation in Arms: A Social Study of the British Army in the First World War*, Manchester University Press, Manchester, 1985, pp. 63-96.

the army insisted that the individuals selected still possessed 'gentlemanly qualities'. Helping to maintain the existing standard were prep-books such as *A General's Letters to His Son*. Its goal was not to teach tactics to the army's newly commissioned junior leaders, but to remind them of the expected behaviour.[152]

It was the army's insistence that leaders epitomise the ideals of a gentleman and adhere to the values of the officer corps that unified its leadership into a monolithic institution. When Lieutenant General Brian Horrocks reminisced on his early days in the army, he remembered that 'we regular army officers of those days might have all come out of the same mould'.[153] The lack of difference was so extreme that for some activities, such as officer appointments, there was an institutional sense that 'one Englishmen is about as good as another' and that selection need not be a difficult or troubling task. Obviously this was an unwise policy but the fact that its suggestion was even possible highlights the extent of the force's uniformity.[154] It was this common social membership and mental outlook that assured the army's inculcation of its ethos, and that gave it a valid alternative to the imposition of doctrine.

In discussing the nature and function of doctrine, this study paper has identified two primary and four secondary characteristics whose presence indicated whether or not a military organisation had become indoctrinated. Except for a few small but highly significant differences, ethos shares these same characteristics. The close fit between the two philosophies underscores the fact that they fulfil similar functions in the wellbeing of a military organisation. They are both vehicles for the advancement of change and the dissemination and inculcation of new ideas, and serve to unify a military force into an intellectual whole. However, despite these similarities, the differences between them are real and important.

Doctrine and ethos share three primary characteristics: both are mechanisms for the learning and dissemination of lessons; both are dynamic, otherwise they risk ossification into dogma; and both embody the expression of a military institution's officially sanctioned ideas. They do not belong to an individual, nor can they exist outside an organisation's corporate mentality.

---

152 See Anon, *A General's Letters to His Son on Obtaining His Commission*, Cassell and Company Ltd., London, 1917.

153 Quoted in Farwell, *Mr Kipling's Army*, p. 145.

154 Ian Hamilton, *A Staff Officer's Scrap-Book During the Russo-Japanese War, Vol. II*, Edward Arnold, London, 1907, p. 318.

Furthermore, in order for ethos to function properly, it must share another characteristic with doctrine. Like doctrine, a military body must embrace its ethos and inculcate it throughout the organisation. In the case of ethos, this is a fairly low threshold to meet. Unlike doctrine, ethos is not an external artefact. Instead, its source lies in the institution's cultural and societal values and, unlike doctrine, it is organic to the organisation.

The critical differences between doctrine and ethos are found in two other significant characteristics. If doctrine is an *intellectual* construct through which a military organisation defines its concept of war and that serves as a vehicle for change, ethos is a *cultural* construct through which a military organisation achieves the same objectives. Whereas a military organisation records its doctrine and disseminates and inculcates it through the publication of official manuals, ethos spreads and is absorbed through a less tangible process of cultural osmosis. The British Army's leaders were aware of the role that culture played in underpinning their army's evolution. At the 1910 conference, during the examination of means to improve the morale of the troops, Colonel J P Du Cane stated that any improvements must be based on the 'idiosyncrasies of our own people' rather than copying the methods of other nations.[155]

The implications underlying the distinctions between doctrine and ethos are far more important than the simple substitution of a few words—intellectual construct versus cultural construct. Doctrine is an imposed, artificial creation, deliberately defined by a military organisation's leadership to meet particular requirements, under particular circumstances, and at particular points in time. Ethos, by contrast, is organic and derives from the spirit of the nation to which a military organisation belongs. Consequently, an ethos-based military organisation more closely reflects, and is more tightly bound to, the values of its parent society. Ethos, therefore, has a natural affinity for its parent society and the two are impossible to separate, except at great risk to both.

The natural affinity between the ethos of a military organisation and the values of its parent society offers another contrast to the more distant relationship that exists between an indoctrinated force and its parent society. To be truly effective doctrine should theoretically conform to societal norms. However, this has not always been the case. For example, during the period under consideration in this study paper, the

155 *Report of a Conference of General Staff Officers at the Staff College 17th to 20th January, 1910*, p. 75.

heavily indoctrinated German officer corps was a separate caste, isolated from the rest of society, and it exerted a political role in national affairs that rivalled the power of the country's elected leaders. In doctrine-based military organisations there is greater risk of a breach between the nation's military and political institutions, and the creation of an imbalance between the responsibilities assigned to the army by its parent society and the roles desired by the force's leaders. A proper civil-military relationship necessitates the maintenance of strong ties between military leaders and their civilian masters. To be truly effective then, doctrine must conform to ethos. Therefore, of these two mechanisms of change, ethos is the more fundamental.

The future of military thinking

For Moltke, the mid-nineteenth century was an era of numerous and pressing social, political and technological demands whose resolution would decide the future course of his nation. He successfully negotiated the military problems that Prussia faced and, in doing so, garnered the maximum benefit from the opportunities available. Moltke's achievements assured that the German people would unite around Prussia, and that the new state of Germany would take its place among the great powers. As a result of his successes, the Prussian Army and its general staff system 'had become the envy of and model for other armed forces both on the continent and beyond. Moltke's military philosophy and strategic approach have survived to inspire successive generations of soldiers.'[156]

It is no surprise that Moltke's innovations were copied by most of the armies of Europe and later the United States. The only major exception was Britain, whose military leaders chose to rely on an institutional ethos, rather than a doctrine, to act as their force's unifying intellectual framework. Yet the strategic decisions that Britain faced were no less complex than those of Prussia. In fact, senior British officers faced challenges that dwarfed the difficulties with which Prussian officers had to contend. Britain's military leaders had to balance responsibility for home defence, the garrisoning and policing of an empire, and the need to prepare and wage war against great power rivals, as well as the incorporation of a host of new technologies and weapon systems. Unlike Prussia, which needed only to consider its problems from the context of its location in central Europe, Britain was a world power with interests, liabilities and threats that spanned the globe.

Contemporary military professionals and defence thinkers, however, must look back on the environment that Prussia and Britain faced with unreserved envy. Since the end of Cold War-induced stability, planners no longer have the 'luxury of focusing primarily upon one set of threat, geographic, and alliance

156 Antulio J Echevarria II, 'Moltke and the German Military Tradition: His Theories and Legacies', *Parameters*, Vol. XXVI, No. 1, Spring 1996, p. 91.

conditions'.[157] As a result, the international environment in which the Western powers operate is more complex, multifaceted, and geographically and technologically diverse than ever before. For example, the extent of the interests that US planners must now strive to balance ranges from the waging of simultaneous conflicts in Iraq, Afghanistan, and the Global War on Terror; the potential for instability in a number of regions including North-East Asia, Latin America and Central Africa; to the possibility of terrorism spreading to new regions, as well as the rapid and unexpected emergence of new threats, even within the US homeland. Their considerations must also incorporate the emergence of Islamic fundamentalism as the antithesis of Western values; the distraction of failed and feeble states, and the ability of organised crime and drug cartels to exploit vulnerable governments; the ongoing requirements of peacekeeping and peace stability; and the admittedly distant but still plausible need to prepare for conventional warfighting scenarios. In addition, US planners are plagued by the need to re-interpret the art of war in light of the warfighting possibilities offered by the Information Revolution in Military Affairs; the economic, military and diplomatic challenges represented by the development of China and a resurgent Russia; the growing political and military risks occasioned by the onset of climate change, and the potential global economic destabilisation resulting from the approaching probability of peak oil. Yet another consideration is the emergence of technologies that have the potential to exacerbate further the threat situation that contemporary military leaders and defence thinkers must face. These include, for example, the weaponisation of biological agents that could provide even the smallest terrorist group with a destructive capability previously the sole province of a major power.[158]

---

157 Gordon R Sullivan and James M Dubik, 'War in the Information Age' in Gordon R Sullivan and James M Dubik (eds), *Envisioning Future War*, US Army Command and General Staff College Press, Fort Leavenworth, 2003, p. 52.

158 John Mackinlay, 'Is UK Doctrine Relevant to Global Insurgency?', *RUSI*, Vol. 152, No. 2, April 2007, p. 36; and Oliver Mintz and Tory House, 'Focused Operations Against Organized Crime in a Mature Peace Operations Environment', *Infantry*, January/February 2007, pp. 38-44; T X Hammes, 'Fourth Generation Warfare Evolves, Fifth Emerges', *Military Review*, Vol. 87, No. 3, May-June 2007, pp. 21-23. For a discussion of the threat of peak oil from a military perspective, see Cameron Leckie, 'Peak Oil and the Australian Army', *Australian Army Journal*, Vol. IV, No. 3, Summer 2007, pp. 21-37.

Moltke saw war as segregated into warfighting and political spheres.[159] Even as late as the Second World War, field commanders could focus their efforts on the defeat of their Axis opponents and treat non-military tasks as a non-core activity. Today's senior officers do not have this luxury. Instead, they must address the political, economic and social factors in their theatre of operations, and not just the application of fire and manoeuvre.[160] As Gordon R Sullivan and James M Dubik observe, there has been a 'blurring of the distinction between "war" and "operations other than war"'.[161]

All of the factors enumerated above suggest that twenty-first century military leaders and defence thinkers will face a threat environment of unprecedented complexity. This environment will require military organisations to stretch the applicability of their doctrines far beyond the bounds of responsiveness that were required in less complex times. There remains one further challenge for the users of doctrine: its ability to react in a timely manner to changes in both the threat environment and the broader evolution in the nature of war.

It will also be difficult to correctly ascertain the most advantageous application of new technologies to the battlefield. Andrew J Bacevich writes that:

> Given the difficulty of assessing such trends—indeed, given the possibility that the phenomena preoccupying us today may be mere blips distracting attention from other deeper currents of change—the proper response to those who claim with certainty to have seen the future of warfare is at least wariness, if not incredulity.[162]

Within the context of the present highly complex and shifting strategic environment and in the face of rapid technological change, it is not unreasonable to question the viability of doctrine, a step that some military thinkers have already taken. When the US Army framed *FM 100-5* in 1976, it was in an era during which the danger was obvious—the Soviet Union. Doctrine writers of that period had a known enemy around which to shape their ideas. Huba Wass de Czege points out that today's theorists do not enjoy

---

159 Hughes, *Moltke on the Art of War*, p. 7.

160 Robert R Leonard, *The Principles of War for the Information Age*, Presidio, Novato, 2000, p. 58.

161 Sullivan and Dubik, 'War in the Information Age', p. 52.

162 A J Bacevich, 'Preserving the Well-Bred Horse', *The National Interest*, Fall 1994, p. 49.

a similar luxury.[163] Rather, he expects, 'given the rate of change in the challenges the US Army will face during this century, it will be impossible to maintain the currency of any method and process based doctrine'.[164] Writing earlier, Jay Luvaas concludes that doctrine might not be sufficient given the range of threats that have emerged since the end of the Second World War. He comments that a doctrine for warfighting focused on one region may not be adequate for somewhere else.[165]

The comments by Wass de Czege and Luvaas would have resonated with the British officers who first considered and then rejected the need for doctrine almost a hundred years ago. Specificity, inflexibility, slowness to adapt, corps provincialism, and a tendency to descend into dogma are among the enduring liabilities of doctrine. This is not surprising. The examples of the interwar armour and air power theorists show that doctrine is often the product of advocates of a pet project rather than the result of deep, well developed argument directed at true insight and transformation. A more recent example lies in the present effort to 'beat the square peg of Iraq into the round hole of COIN doctrine—as Robert Thompson and David Galula annunciated it for an entirely different situation four decades ago'.[166] The British Lieutenant General John Kiszely has similarly warned of the tendency of doctrine seekers to 'fit the circumstances to the doctrine, rather than the other way round—thereby trying to turn it into something that is alien to its nature'.[167]

Another problem with doctrine is a tendency for platitudinous declarations, and contemporary manuals are filled with trite phrases that bear a striking resemblance to those that afflicted the pages of the British Army's *Field Service Regulations*. Williamson Murray scathingly describes some recent US doctrine publications as 'mind-numbing prose', lacking in intellectual content and containing 'interminable

---

163 Huba Wass de Czege, 'Lessons from the Past: Getting the Army's Doctrine "Right Enough" Today' in Scott Hopkins (ed), *2006 Chief of Army's Exercise Proceedings*, Land Warfare Studies Centre, Canberra, 2007, p. 82.

164 Ibid., p. 101.

165 Luvaas, 'Some Vagrant Thoughts on Doctrine', p. 59.

166 Jim Lacey and Kevin Woods, 'Adapt or Die', *United States Naval Institute Proceedings*, Vol. 133, No. 8, p. 19. See also David Galula, *Counterinsurgency Warfare: Theory and Practice*, Praeger, Westport, 2005; and Robert Thompson, *Defeating Communist Insurgency: Experiences from Malaya and Vietnam*, Praeger, Westport, 1966.

167 John Kiszely, 'Learning about Counter-Insurgency', *RUSI Journal*, Vol. 151, No. 6, December 2006, p. 19.

laundry lists of bureaucratic concerns'. He concludes that they are 'harmless, except to those condemned to read them'.[168] Luvaas, in his seminal work on military learning and the American Civil War, observes that doctrine can be either 'servant or master', and that it can 'narrow a soldier's vision by dictating the questions and forming the basis for judgement'.[169]

But even these observations on the liabilities and limitations of doctrine do not go far enough. The powerful institutional pull of doctrine in the contemporary military mindset is such that officers and theorists alike are unable to divorce themselves completely from the concept. James Corum, who is particularly critical of US comprehension of the requirements for waging war in Iraq, believes that part of the solution to the crisis in Baghdad lies in the crafting of a more appropriate doctrine.[170] Gordon and Dubik, who expect the Information Age to fundamentally 'change army organizations, processes and operations—as well as the conduct of war',[171] still look to doctrine as the means of achieving these advancements.[172] Even Robert R Leonhard, who offers *The Principles of War for the Information Age* as a direct challenge to the US Army's core beliefs, cannot bring himself to question doctrine. Leonhard's call for the army to 'smash a few idols' stops short of an abandonment of the hitherto sacrosanct doctrine.[173]

When the British considered and rejected indoctrination in the early years of the twentieth century they did so because they believed it was inappropriate to their requirements. Instead, they decided to continue to rely on their ethos as their guide. The US military today must examine the same option, as must the military organisations of other Western powers as they confront the challenges of the twenty-first century. Military leaders and defence thinkers already recognise that radical change is necessary

---

168 Williamson Murray, 'Does Military Culture Matter?', *Orbis*, Winter 1999, p. 40.

169 Luvaas, *The Military Legacy of the Civil War*, p. xviii.

170 James S Corum, *Fighting the War on Terror: A Counterinsurgency Strategy*, Zenith Press, St Paul, MN, 2007, p. 242.

171 Gordon and Dubik, 'War in the Information Age', p. 54.

172 Ibid., p. 58.

173 Leonhard, *The Principles of War in the Information Age*, p. xiv.

if they are to exploit the potential of the Information Revolution in Military Affairs, and they acknowledge the necessity of unlearning the rules of Industrial Age warfare.[174]

Yet the replacement of doctrine with another intellectual mechanism will be no easy task. Military organisations are imbued with culturally biased preferences of which doctrine is just one. Culture is deeply imbedded in the fabric of military institutions and its tenets are vigorously defended by traditionalists who resist modification at all costs. Consequently, as Murray points out, cultural change may take decades to effect.[175] However, institutional resistance must be overcome, and the sooner the process begins, the sooner a new mechanism will be found.

It would be tempting to simply assert the validity of British 'institutional ethos' and call for its recreation for the present. This would be a mistake. Like doctrine, the British Army's ethos was the product of a particular time and a reaction to a particular set of circumstances and challenges. The British officer corps that used ethos was an insular body whose members were largely drawn from the same class and who shared similar outlooks and opinions. The British Army of today is a different institution from that which relied on ethos in the era of the First World War. As with doctrine, the time of British ethos has passed.

Moreover, to recreate an ethos or redefine a doctrine would violate one of the key observations on Revolutions in Military Affairs—the importance of radical change. The Information Revolution will be no exception.[176] While it may not yet be possible to discern the exact nature of the Information Age, it is time to begin the process of identifying a new intellectual mechanism that will help military organisations adapt to this latest Revolution in Military Affairs. What is required is to begin the debate on doctrine's suitability and the nature of its replacement. This will require more than tinkering around the edges or the consideration of minor modifications. Rather, it will require institutions to undertake vigorous and rigorous debate, encourage deep thinking, permit independent thought and allow unimpeded argument. Military organisations must be willing to abandon their comfort zones and step forward into the unknown.

---

174 Gordon and Dubik, 'War in the Information Age', p. 50. For an early, but still highly relevant work on this theme, see Eliot A Cohen, 'A Revolution in Warfare', *Foreign Affairs*, Vol. 75, No. 2, March/April 1996, pp. 37-54.

175 Murray, 'Does Military Culture Matter?', p. 28.

176 See Thomas G Mahnken, 'Transforming the US Armed Forces: Rhetoric or Reality?', *Naval War College Review*, Vol. LIV, No. 3, Summer 2001, pp. 85-99.

Conclusion

Doctrine developed when the dominant form of war was state-on-state conflict waged between similarly armed, organised and trained conventional forces. The Information Revolution in Military Affairs has the ability to usher in a new age of warfare in which the possibility of direct inter-state conflict has receded and the threat of 'asymmetric warfare' has grown in prominence. Much progress has been made by the United States, for example, in exploiting the military potential of new technologies. However, during the current Iraq War the United States has again demonstrated the liability of planning for battle in accordance with a particular doctrine. As in the Vietnam War, the United States in Iraq employed a concept of warfighting that was severely disconnected from the requirements of the conflict. Vietnam ended in a conclusive defeat while in Iraq a final decision awaits.

In Vietnam the United States attempted to wage the war it wanted rather than 'the war it got'. By insisting on a conventional warfare strategy (designed to fight the Soviet Union in Europe) against a Maoist insurgency, the United States doomed itself to defeat in South-East Asia.[177] In Iraq, a new generation of US military leaders, convinced of the superiority of their high-tech forces, found themselves ensnared in a war that they were ill-prepared to fight once the opening conventional phase ended. In both conflicts, the United States did not accurately anticipate the nature of the task.

The United States' intellectual response to the problems confronting it in Iraq is troubling. General David Petraeus has admitted that: 'The insurgencies in Iraq and Afghanistan were not, in truth, the wars for which we were best prepared in 2001; however, they are the wars we are fighting and they clearly are the kind of wars we must master.'[178] Yet it was only after several years of witnessing a worsening situation in the Middle East, and after strident calls for a change of approach, that the United

177 Andrew F Krepinevich, Jr, *The Army and Vietnam*, Johns Hopkins University Press, Baltimore, 1989, pp. 4-5.

178 David H Petraeus, 'Learning Counterinsurgency: Observations from Soldering in Iraq', *Military Review*, Vol. 86, No. 1, January/February 2006, p. 2.

States modified its concept of war for Iraq.[179] It was not until early 2007, after Petraeus became Commander of the Multi-National Force-Iraq, that the United States committed additional forces as part of a troop surge and transitioned to a strategy that incorporated the principles of counterinsurgency rather than merely those of conventional war.

It is too soon to tell whether the change in strategy that Petraeus implemented will achieve the desired effect, especially since the United States still lacks a 'whole-of-government' plan. What is more worrying, however, is the US Army's remarkable metamorphosis in which its forces easily and freely substituted doctrines. If, as its adherents assert, doctrine is the core belief system of military organisations, the expectation that a force can readily implement dramatic, even oppositional, changes in direction is more than a little puzzling. The depth of the US Army's embrace of counterinsurgency warfare is particularly suspect when it has not been accompanied by reforms to organisation, force structure, and weapon mix.

Even after the passage of a year it is by no means apparent that Petraeus has implemented an appropriate counterinsurgency doctrine. A number of observers have expressed their concern that the new approach in Iraq is fundamentally flawed because it derives from an incorrect application of the lessons of history. What the US has done is apply to Iraq the classic counterinsurgency tenets that the British Army developed for Malaya during the Emergency. The problem with this transference lies in the basic differences that exist between the two insurgencies. The Malayan Emergency was a protracted guerrilla war guided by a Maoist communist ideology and motivated by nationalist objectives. In Iraq the drivers fuelling the conflict are religious, ethnic and tribal, underpinned by armed militias and organised crime, with a degree of participation by al-Qaeda operatives.[180] The situations are not the same, and hence neither will be the solutions. In seeking a quick fix for its inappropriate application of its doctrine of conventional war, the United States has embraced a different doctrine without pausing to understand the nature of the conflict which it

---

179 Corum, *Fighting the War on Terror*, p. 266.

180 Warren Chin, 'Examining the Application of British Counterinsurgency Doctrine by the American Army in Iraq', *Small Wars and Insurgencies*, Vol. 18, No. 1, March 2007, p. 14; and Steven Metz, *Rethinking Insurgency*, Strategic Studies Institute, Carlisle, 2007, pp. 42-43.

has to fight. In effect, the United States has applied an 'outdated and even obsolete doctrine, not appropriate to the current situation it faces in Iraq'.[181]

Stephen Biddle also notes the false linkage between Iraq and Vietnam. He insists that 'the current struggle is not a Maoist "people's war" of national liberation; it is a communal civil war with very different dynamics'.[182] Nor, as Metz observes, should the Viet Cong be treated as 'the archetypical foe'.[183] Biddle believes that 'recycling the Vietnam playbook' will actually make the situation in Iraq worse, and that it will prove counterproductive to implement a strategy designed to win 'hearts and minds' in a society that is increasingly polarised.[184]

Only the passage of time will reveal whether Petraeus has found the 'solution' to Iraq. But whether victory, defeat, or some form of chaos in between is the future of the US presence in the Middle East, this must not obscure the fact that the war in Iraq is one type of conflict, involving distinct cultures and peoples, taking place in a particular location, and occurring during a certain time period. It is a unique historical event whose requirements are unlikely to be satisfied by an off-the-shelf doctrine.

In fact, new doctrine may not be what is needed for the new wars of the twenty-first century. The example of the British Army in the era of the First World War demonstrates that doctrine is not as essential as contemporary military professionals and defence thinkers would like to believe. Nor is doctrine, as the British employment of institutional ethos indicates, the only viable mechanism with which to interpret the challenges of war and through which to identify, disseminate and inculcate solutions.

The British Army experience of ethos provides lessons for all military organisations as they struggle to find solutions to today's increasingly difficult threat environment. Complex and geographically diverse situations demand mechanisms for flexible response and innovation. By leaping from one form of doctrine to another, the United States may secure victory in Iraq. Equally, however, it may set the foundation for defeat in its next conflict. Military organisations equipped with ill-conceived doctrinal models

---

181 Chin, 'Examining the Application of British Counterinsurgency Doctrine by the American Army in Iraq', p. 14.

182 Stephen Biddle, 'Seeing Baghdad, Thinking Saigon', *Foreign Affairs*, Vol. 85, No. 2, March/April 2006, p. 2.

183 Steven Metz, 'New Challenges and Old Concepts: Understanding 21st Century Insurgency', *Parameters*, Vol. XXXVII, No. 4, Winter 2007-08, p. 21.

184 Biddle, 'Seeing Baghdad, Thinking Saigon', p. 7.

will find themselves embroiled in conflicts that they are unsuited to wage. To attempt to regulate the future of war through doctrine, in the face of the vast array of variables that confronts today's military professionals, is to commit intellectual hubris. Such efforts would amount to nothing more than attempting to confine war within known and comfortable boundaries, a policy that, as the French discovered in 1940, can only lead to disaster.

It is a truism of war that military organisations are not always able to choose the nature of the conflicts they must fight. The experiences of France in 1940, the United States in Vietnam and now again in Iraq, underscore the danger military organisations face when they limit themselves to a single concept of war. The risks of an adherence to doctrine may be too great for the present and future conflicts of the twenty-first century. Doctrine flourished in an age of specificity and obvious threat. That age has clearly passed. The priority now is for military organisations to define new mechanisms to guide them through the challenges of the future.

Bibliography

Unpublished Sources

Imperial War Museum

Christison Collection
Ironside Collection
Maxse Collection
Murray Collection

Liddell Hart Centre for Military Archives

Adam Collection
Burnett-Stuart Collection
Capper Collection
Shea Collection

Published Sources

Anon, 'The British Army and Modern Conceptions of War', *The Edinburgh Review*, Vol. 213, No. 436, April 1911; and reprinted in *Journal of the Royal United Service Institution*, Vol. 55, September 1911.

Anon, 'The Place of Doctrine in War', *The Edinburgh Review*, Vol. 215, No. 439, January 1912.

Anon, 'The Staff', *The Army Quarterly*, Vol. 1, No. 1, October, 1920.

Anon, 'The Development of the General Staff', *The Army Review*, Vol. 1, No. 1, July 1911.

Anon, 'The General Staff in India', *The Army Review*, Vol. 2, No. 1, January 1912.

Anon, *A General's Letters to His Son on Obtaining His Commission*, Cassell and Company Ltd, London, 1917.

Australian Army, *Land Warfare Doctrine 1 – Fundamentals of Land Warfare*, Land Warfare Development Centre, Puckapunyal, 2002.

Australian Army, *The Fundamentals of Land Warfare*, CDTC, Canberra, 1998.

Bacevich, A J, 'Preserving the Well-Bred Horse', *The National Interest*, Fall 1994.

Battine, C W, 'The Proposed Changes in Cavalry Tactics', *Journal of the Royal United Service Institution*, Vol. 54, No. 393, November 1910.

Biddle, Stephen, 'Seeing Baghdad, Thinking Saigon,' *Foreign Affairs*, Vol. 85, No. 2, March/April 2006.

Bidwell, Shelford and Dominick Graham, *Fire-Power: British Army Weapons and Theories of War 1904-1945*, George Allen & Unwin, London, 1982.

Black, Jeremy, *A Military History of Britain: From 1775 to the Present*, Praeger Security International, Westport, 2006.

Bond, Brian, 'Doctrine and Training in the British Cavalry, 1870-1914' in Michael Howard (ed), *The Theory and Practice of War: Essays Presented to Captain B H Liddell Hart*, Oxford, London, 1965.

Bond, Brian, *The Victorian Army and the Staff College, 1854-1914*, Eyre Methuen, London, 1972.

Bond, Brian and Williamson Murray, 'The British Armed Forces, 1918-39' in Allan R Millett and Williamson Murray (eds), *Military Effectiveness, Volume II: The Interwar Period*, Allen & Unwin, London, 1988.

Bourne, J M, 'Haig and the Historians' in Brian Bond and Nigel Cave (eds), *Haig: A Reappraisal 70 Years On*, Leo Cooper, Barnsley, 1999.

Bucholz, Arden, *Moltke, Schlieffen and Prussian War Planning,* Berg, New York, 1991.

Bucholz, Arden, *Moltke and the German Wars, 1864-1871*, Palgrave, Houndmills, Basingstoke, 2001.

Childs, David J, *A Peripheral Weapon?: The Production and Employment of British Tanks in the First World War*, Greenwood Press, Westport, 1999.

Chin, Warren, 'Examining the Application of British Counterinsurgency Doctrine by the American Army in Iraq', *Small Wars and Insurgencies*, Vol. 18, No. 1, March 2007.

Citino, Robert M, *Quest for Decisive Victory: From Stalemate to Blitzkrieg in Europe, 1899-1940*, University Press of Kansas, Lawrence, 2002.

von Clausewitz, Carl, *On War*, Michael Howard and Peter Paret (eds and trans), Princeton University Press, Princeton, 1989.

Cohen, Eliot A, 'A Revolution in Warfare', *Foreign Affairs*, Vol. 75, No. 2, March/April 1996.

Coop, J O, *The Story of the 55th (West Lancashire Division)*, Daily Post, Liverpool, 1919.

Corum, James S, *Fighting the War on Terror: A Counterinsurgency Strategy*, Zenith Press, St Paul, 2007.

Department of the Army, *FM 3-0 Operations*, Washington DC, 2001.

Doughty, Robert Allan, *The Seeds of Disaster: The Development of French Army Doctrine, 1919-1939*, Archon Books, Hamden, 1985.

Echevarria II, Antulio J, 'Moltke and the German Military Tradition: His Theories and Legacies', *Parameters*, Vol. XXVI, No. 1, Spring 1996.

English, John A, *The Canadian Army and the Normandy Campaign: A Study in Failure in High Command*, Praeger, Westport, 1991.

English, John A, *Marching Through Chaos: The Descent of Armies in Theory and Practice*, Praeger, Westport, 1996.

Evans, Michael, *Forward from the Past: The Development of Australian Doctrine 1972 – Present*, Land Warfare Studies Centre, Canberra, 1999.

Farwell, Byron, *Mr. Kipling's Army: All the Queen's Men*, Norton and Company, New York, 1981.

Franks, Jr, Frederick M, 'Continuity and Change: Discussing Our Evolving Doctrine', *Military Review*, Vol. LXXI, No. 10, October 1991.

French, David, *Raising Churchill's Army: The British Army and the War against Germany 1919-1945*, Oxford University Press, Oxford, 2000.

Fuller, J F C, *Training Soldiers for War*, Hugh Rees Ltd, London, 1914.

Fuller, J F C, 'The Tactics of Penetration: A Counterblast to German Numerical Superiority', *Journal of the Royal United Service Institution*, Vol. 59, November 1914.

Fuller, J F C, *The Foundations of the Science of War*, Hutchinson & Co, London, 1925.

Fuller, J F C, *Memoirs of an Unconventional Soldier*, Ivor Nicholson and Watson, London, 1936.

The Fundamentals of British Maritime Doctrine, 1996.

Galula, David, *Counterinsurgency Warfare: Theory and Practice*, Praeger, Westport, 2005.

General Staff, War Office, *Field Service Regulations, Part I, Operations*, HMSO, London, 1909.

General Staff, War Office, *Field Service Regulations, Volume II, Operations – General*, HMSO, London, 1935.

General Staff, War Office, *Field Service Regulations, Volume II, Operations (Provisional)*, HMSO, London, 1920.

General Staff, War Office, *Mechanized and Armoured Formations*, HMSO, London, 1929.

General Staff, War Office, *Report of a Conference of General Staff Officers at the Staff College, 17th to 20th January 1910*, HMSO, London, 1910.

General Staff, War Office, *Report of a Conference of General Staff Officers at the Staff College, 9th to 12th January, 1911*, HMSO, London, 1911.

Graham, DS, 'The Ascendancy of Firepower' in JA English, J Addicott and PJ Kramers (eds), *The Mechanized Battlefield: A Tactical Analysis,* Pergamon-Brassey's, Washington, 1985.

Graham, Dominick, 'Sans Doctrine: British Army Tactics in the First World War' in Timothy Travers and Christon Archer (eds), *Men at War: Politics, Technology and Innovation in the Twentieth Century*, Precedent, Chicago, 1982.

Graham, Dominick, 'Observations on the Dialectics of British Tactics, 1904-45', in Ronald Haycock and Keith Neilson (eds), *Men, Machines & War*, Wilfrid Laurier University Press, Waterloo, 1988.

Grant, Charles, 'The Use of History in the Development of Contemporary Doctrine' in John Gooch (ed), *The Origins of Contemporary Doctrine*, Strategic & Combat Studies Institute, Camberley, 1997.

Great Britain, General Staff, *Design for Military Operations – The British Military Doctrine*, 1989.

Griffith, Paddy, *Battle Tactics of the Western Front: The British Army's Art of Attack, 1916-1918*, Yale University Press, New Haven, 2004.

Gudmundsson, Bruce I, *Stormtroop Tactics: Innovation in the German Army, 1914-1918*, Praeger, New York, 1989.

Hamilton, Ian, *A Staff Officer's Scrap-Book During the Russo-Japanese War, Vol. II*, Edward Arnold, London, 1907.

Hammes, T X, 'Fourth Generation Warfare Evolves, Fifth Emerges', *Military Review*, Vol. 87, No. 3, May/June 2007.

Harris, JP, *Men, Ideas and Tanks: British Military Thought and Armoured Forces, 1903-1939*, Manchester University Press, Manchester, 1995.

Harris, J P, with Niall Barr, *Amiens to the Armistice: The BEF in the Hundred Days' Campaign, 8 August – 11 November 1918,* Brassey's, London, 1998.

Hart, Russell A, *Clash of Army: How the Allies Won in Normandy*, Lynne Rienner, Boulder, 2001.

Hart, Stephen Ashley, *Montgomery and 'Colossal Cracks': The 21st Army Group in Northwest Europe, 1944-45*, Praeger, Westport, 2000.

Henderson, G F R, 'Tactics of the Three Arms Combined', reprinted in G F R Henderson, *The Science of War: A Collection of Essay and Lectures 1891-1903*, Longmans, Green and Co, London, 1910.

Herbert, Paul H, *Deciding What Has to Be Done: General William E DePuy and the 1976 Edition of FM 100-5, Operations*, Combat Studies Institute, Fort Leavenworth, 1988.

Holborn, Hajo, 'The Prusso-German School: Moltke and the Rise of the General Staff' in Peter Paret (ed), *Makers of Modern Strategy from Machiavelli to the Nuclear Age*, Princeton University Press, Princeton, 1986.

Holden Reid, Brian, 'Introduction: Is There a British Military Doctrine' in J J G Mackenzie and Brian Holden Reid (eds), *Central Region vs Out-of-Area: Future Commitments*, Tri Service Press, Camberley, 1990.

Holden Reid, Brian, *War Studies at the Staff College, 1890-1930*, Strategic & Combat Studies Institute, Camberley, 1992.

Holden Reid, Brian, *A Doctrinal Perspective, 1988-98*, Strategic & Combat Studies Institute, Camberley, 1998.

Howard, Michael, *The Franco-Prussian War: The German Invasion of France, 1870-1871*, Collins, London, 1967.

Johnson, Garry, 'On Winning: An Option for Change Without Decay', *The Journal of the Royal United Service Institution*, Autumn, 1991.

Jones, Archer, *The Art of War in the Western World*, Oxford University Press, Oxford, 1989.

Kiszely, John, 'Learning about Counter-Insurgency', *RUSI Journal*, Vol. 151, No. 6, December 2006.

Kiszely, John, *Post-Modern Challenges for Modern Warriors*, The Shrivenham Papers No. 5, Defence Academy of the United Kingdom, Shrivenham, 2007.

Kochanski, Halik, 'Planning for War in the Final Years of *Pax Britannica*, 1889-1903', in David French and Brian Holden Reid (eds), *The British General Staff: Reform and Innovation c.1890-1939*, Frank Cass, London, 2002.

Krepinevich, Jr., Andrew, *The Army and Vietnam*, Johns Hopkins University Press, Baltimore, 1989.

Lacey, Jim and Kevin Woods, 'Adapt or Die', *United States Naval Institute Proceedings*, Vol. 133, No. 8, 2007.

Langlois, H, *Lessons from Two Recent Wars (The Russo-Turkish and South African Wars)*, HMSO, London, 1909.

Leckie, Cameron, 'Peak Oil and the Australian Army', *Australian Army Journal*, Vol. IV, No. 3, Summer 2007.

Leonard, Robert R, *The Principles of War for the Information Age*, Presidio, Novato, 2000.

Low, T H, 'Discussion on a Plea for a Mission and a Doctrine', *The Marine Corps Gazette*, Vol. I, No. 3, September 1916.

Lupfer, Timothy T, *The Dynamics of Doctrine: The Changes in German Tactical Doctrine During the First World War*, Combat Studies Institute, Fort Leavenworth, 1981.

Luvaas, Jay, 'Some Vagrant Thoughts on Doctrine', *Military Review*, March 1986.

Luvaas, Jay, *The Military Legacy of the Civil War: The European Inheritance*, University Press of Kansas, Lawrence, 1988.

Mackinlay, John, 'Is UK Doctrine Relevant to Global Insurgency?', *RUSI Journal*, Vol. 152, No. 2, April 2007.

Mahnken, Thomas G, 'Transforming the US Armed Forces: Rhetoric or Reality?', *Naval War College Review*, Vol. LIV, No. 3, Summer 2001.

Margiotta, Franklin D, *Brassey's Encyclopedia of Land Forces and Warfare*, Brassey's, Washington DC, 2000.

McAnally, J H S, 'The purposes and benefits of doctrine: Why go to all the trouble of having one?' in Michael Duffy, Theo Farrell & Geoffrey Sloan (eds), *Doctrine and Military Effectiveness: Proceedings of the Conference held at the Britannia Royal Navy College, January 16th-17th, 1997*, The Strategic Policy Studies Group of the Britannia Royal Naval College & Exeter University, Exeter, 1997.

McDonough, James R, 'Building the New FM 100-5: Process and Product', *Military Review*, Vol. LXXI, No. 10, October 1991.

Metz, Steven, *Rethinking Insurgency*, Strategic Studies Institute, Carlisle, 2007.

Metz, Steven, 'New Challenges and Old Concepts: Understanding 21st Century Insurgency', *Parameters*, Vol. XXXVII, No. 4, Winter 2007-08.

Mintz, Oliver and Tory House, 'Focused Operations Against Organized Crime in a Mature Peace Operations Environment', *Infantry*, Vol. 96, Iss. 1, January/February 2007.

Moreman, T R, *The Army in India and the Development of Frontier Warfare, 1849-1947*, Palgrave, London, 1998.

'Mousquetaire', 'Heretical Doctrines of War', *United Service Magazine*, Vol. 66, January 1913.

Murray, Williamson, 'Does Military Culture Matter?', *Orbis*, Winter 1999.

Musgrave, H, 'Various Aspects of Preparation for War', *The Army Review*, Vol. 5, July 1913.

Nagl, John A, *Learning to Eat Soup with a Knife: Counterinsurgency Lessons for Malaya and Vietnam*, The University of Chicago Press, Chicago, 2005.

Odom, William O, *After the Trenches: The Transformation of US Army Doctrine, 1918-1939*, Texas A&M University Press, College Station, 1999.

Overy, Richard, 'Doctrine Not Dogma: Lessons from the Past' in Michael Duffy, Theo Farrell & Geoffrey Sloan (eds), *Doctrine and Military Effectiveness: Proceedings of the Conference held at the Britannia Royal Navy College, January 16th-17th, 1997*, The Strategic Policy Studies Group of the Britannia Royal Naval College & Exeter University, Exeter, 1997.

Palazzo, Albert, 'The British Army's Counter-Battery Staff Office and Control of the Enemy in World War I', *The Journal of Military History*, Vol. 63, No. 1, January 1999.

Palazzo, Albert, *Seeking Victory on the Western Front: The British Army and Chemical Warfare in World War I*, University of Nebraska Press, Lincoln, 2000.

Petraeus, David H, 'Learning Counterinsurgency: Observations from Soldering in Iraq', *Military Review*, Vol. 86, No. 1, January/February 2006.

Pollack, AWA, 'Strategy in Theory and Practice', *United Service Magazine*, Vol. 67, July 1913.

Ramsay, M A, *Command and Cohesion: The Citizen Soldier and Minor Tactics in the British Army, 1870-1918*, Praeger, Westport, 2002.

Rosen, Stephen Peter, *Winning the Next War: Innovation and the Modern Military*, Cornell University Press, Ithaca, 1991.

Rothenberg, Gunther E, 'Moltke, Schlieffen, and the Doctrine of Strategic Envelopment' in Peter Paret (ed), *Makers of Modern Strategy from Machiavelli to the Nuclear Age*, Princeton University Press, Princeton, 1986.

Rothenberg, Gunther E, 'Forward' in Daniel J Hughes (ed), Daniel J Hughes and Harry Bell (trans), *Moltke on the Art of War: Selected Writings*, Presidio Press, Novato, 1993.

Russell, John H, 'A Plea for a Mission and a Doctrine', *The Marine Corps Gazette*, Vol. I, No. 2, June 1916.

Sheffield, G D, 'The Indispensable Factor: The Performance of British Troops in 1918' in Peter Dennis and Jeffrey Grey (eds), *Defining Victory, 1918*, Army History Unit, Canberra, 1999.

Showalter, Dennis E, *Railroads and Rifles: Soldiers, Technology and the Unification of Germany*, Archon Books, Hamden, 1975.

Showalter, Dennis E, 'The Prusso-German RMA, 1840-1871' in Macgregor Knox and Williamson Murray (eds), *The Dynamics of Military Revolution 1300-2050*, Cambridge University Press, Cambridge, 2001.

Simpson, Keith, 'The Officers' in Ian F W Beckett and Keith Simpson, *A Nation in Arms: A Social Study of the British Army in the First World War*, Manchester University Press, Manchester, 1985.

Snyder, Jack, *The Ideology of the Offensive: Military Decision Making and the Disasters of 1914*, Cornell University Press, Ithaca, 1984.

Spiers, Edward M, *Haldane: An Army Reformer*, Edinburg University Press, Edinburg, 1980.

Spiller, Roger J, 'In the Shadow of the Dragon: Doctrine and the US Army after Vietnam' in Jeffrey Grey and Peter Dennis (eds), *From Past to Future: The Australian Experience of Land/Air Operations*, Australian Defence Force Academy, Canberra, 1995.

Starry, Donn A, 'Changing an Army: An American Perspective' in Michael Evans (ed), *Changing the Army: The Roles of Doctrine, Development and Training*, Land Warfare Studies Centre, Duntroon, 2000.

Stone, John, 'The British Army and the Tank' in Theo Farrell and Terry Terriff (eds), *The Sources of Military Change: Culture, Politics, Technology*, Lynne Rienner Publishers, Boulder, 2002.

Strachan, Hew, 'The British Way in Warfare' in David Chandler (ed), *The Oxford Illustrated History of the British Army*, Oxford University Press, Oxford, 1994.

Sullivan, Gordon R, 'Doctrine: A Guide to the Future', *Military Review*, Vol. LXXII, No. 2, February 1992.

Sullivan, Gordon R and James M Dubik, 'War in the Information Age' in Gordon R Sullivan and James M Dubik (eds), *Envisioning Future War*, US Army Command and General Staff College Press, Fort Leavenworth, 2003.

Thompson, Robert, *Defeating Communist Insurgency: Experiences from Malaya and Vietnam*, Praeger, Westport, 1966.

Travers, Tim, 'Learning and Decision-Making on the Western Front, 1915-1916: The British Example', *Canadian Journal of History*, Vol. 18, No. 1, 1983.

Travers, Tim, *The Killing Ground: The British Army, The Western Front and the Emergence of Modern Warfare, 1900-1918*, Unwin Hyman, London, 1987.

Travers, Tim, *How the War Was Won: Command and Technology in the British Army on the Western Front 1917-1918*, Routledge, London, 1992.

Tuker, Francis, *Approach to Battle: A Commentary, Eighth Army – November 1941 to May 1943*, Cassell, London, 1963.

'Veto', 'Is 'Doctrine' a Delusion?', *United Service Magazine*, Vol. 66, November 1912.

War Office, *Notes on Elementary Tactical Training*, HMSO, London, 1923.

Wass de Czege, Huba, 'Lessons from the Past: Getting the Army's Doctrine "Right Enough" Today' in Scott Hopkins (ed), *2006 Chief of Army's Exercise Proceedings*, Land Warfare Studies Centre, Canberra, 2007.

Weber, W H F, *A Field Artillery Group in Battle: A Tactical Study Based on the Actions of 2nd Brigade, R F A during the German Offensive, 1918, the 100 Day's Battle, and the Battle of Cambrai, 1917*, Royal Artillery Institute, London, 1919.

Wilkinson, Spencer, *The Brain of an Army*, Greg Revivals, Aldershot, 1992.

Winton, Harold, *To Change an Army: General Sir John Burnett-Stuart and British Armored Doctrine, 1927-1938*, University Press of Kansas, Lawrence, 1988.

Wise, S F, 'The Black Day of the German Army: Australian and Canadians at Amiens, August 1918' in Peter Dennis and Jeffrey Grey (eds.), *Defining Victory, 1918*, Army History Unit, Canberra, 1999.

Land Warfare Studies Centre

Publications

The General Sir Brudenell White Monograph Series

Schmidtchen, David, *The Rise of the Strategic Private: Technology, Control and Change in a Network-Enabled Military*, 2006.

Study Papers

301 Evans, Michael, *Forward from the Past: The Development of Australian Army Doctrine, 1972-Present*, August 1999.

302 Ryan, Alan, *From Desert Storm to East Timor: Australia, the Asia-Pacific and the 'New Age' Coalition Operations*, January 2000.

303 Evans, Michael, *Developing Australia's Maritime Concept of Strategy: Lessons from the Ambon Disaster of 1942*, April 2000.

304 Ryan, Alan, *'Primary Responsibilities and Primary Risks': Australian Defence Force Participation in the International Force East Timor*, November 2000.

305 Evans, Michael, *The Continental School of Strategy: The Past, Present and Future of Land Power, June 2004.*

306 Evans, Michael, *The Tyranny of Dissonance: Australia's Strategic Culture and Way of War, 1901-2005* , February 2005.

307 Morrison, James C, *Mechanising an Army: Mechanisation and the Conversion of the Light Horse, 1920-1943*, June 2006.

308 Hopkins, Scott (ed), *Asymmetry and Complexity: Selected Papers from the 2005 Rowell Seminar and the 2005 Chief of Army's Conference*, February 2007.

309 Hopkins, Scott (ed), *2006 Chief of Army's Exercise Proceedings*, March 2007.

310 Wegener, Andrew, *A Complex and Changing Dynamic: Afghan Responses to Foreign Intervention, 1878-2006*, April 2007.

311 Stockings, Craig, *The Making and Breaking of the Post-Federation Australian Army, 1901-09*, July 2007.

312 Keating, Gavin, *A Tale of Three Battalions: Combat Morale and Battle Fatigue in the 7th Australian Infantry Brigade, Bougainville, 1944-45*, September 2007.

313 *Chief of Army's Reading List*, December 2007.

314 Breen, Bob and Greg McCauley, *The World Looking Over Their Shoulders: Australian Strategic Corporals on Operations in Somalia and East Timor*, August 2008.

Working Papers

101 Evans, Michael, *The Role of the Australian Army in a Maritime Concept of Strategy*, September 1998.

102 Dunn, Martin, *Redefining Strategic Strike: The Strike Role and the Australian Army into the 21st Century*, April 1999.

103 Evans, Michael, *Conventional Deterrence in the Australian Strategic Context*, May 1999.

104 de Somer, Greg, *The Implications of the United States Army's Army-After-Next Concepts for the Australian Army*, June 1999.

105 Ryan, Alan, *The Way Ahead? Alternative Approaches to Integrating the Reserves in 'Total Force' Planning*, July 1999.

106 de Somer, Greg, *The Capacity of the Australian Army to Conduct and Sustain Land Force Operations*, August 1999, reprinted October 1999.

107 de Somer, Greg and David Schmidtchen, *Professional Mastery: The Human Dimension of Warfighting Capability for the Army-After-Next*, October 1999.

108 Zhou, Bo, *South Asia: The Prospect of Nuclear Disarmament After the 1998 Nuclear Tests in India and Pakistan*, November 1999.

109 Ryan, Michael and Michael Frater, *A Tactical Communications System for Future Land Warfare*, March 2000.

110 Evans, Michael, *From Legend to Learning: Gallipoli and the Military Revolution of World War I*, April 2000.

111 Wing, Ian, *Refocusing Concepts of Security: The Convergence of Military and Non-military Tasks*, November 2000.

112 Ryan, Michael and Michael Frater, *The Utility of a Tactical Airborne Communications Subsystem in Support of Future Land Warfare*, April 2001.

113 Evans, Michael, *From Deakin to Dibb: The Army and the Making of Australian Strategy in the 20th Century*, June 2001.

114 Ryan, Alan, *Thinking Across Time: Concurrent Historical Analysis on Military Operations*, July 2001.

115 Evans, Michael, *Australia and the Revolution in Military Affairs*, August 2001.

116 Frater, Michael and Michael Ryan, *Communications Electronic Warfare and the Digitised Battlefield*, October 2001.

117 Parkin, Russell, *A Capability of First Resort: Amphibious Operations and the Australian Defence Policy, 1901-2001*, May 2002.

118 Blaxland, John, *Information-era Manoeuvre: The Australian-led Mission to East Timor*, June 2002.

119 Connery, David, *GBAeD 2030: A Concept for Ground-based Aerospace Defence in the Army-After-Next*, July 2002.

120 Beasley, Kent, *Information Operations during Operation Stabilise in East Timor*, August 2002.

121 Ryan, Alan, *Australian Army Cooperation with the Land Forces of the United States: Problems of the Junior Partner*, January 2003.

122 Evans, Michael and Alan Ryan (eds), *From Breitenfeld to Baghdad: Perspectives on Combined Arms Warfare*, January 2003.

123 Hoare, Mark, *The Prospects for Australian and Japanese Security Cooperation in a More Uncertain Asia-Pacific*, July 2003.

124 Ryan, Alan, *'Putting Your Young Men in the Mud': Change, Continuity and the Australian Infantry Battalion*, September 2003.

125 Schmidtchen, David, *Network-Centric Warfare: The Problem of Social Order*, July 2005.

126 Watson, James, *A Model Pacific Solution? A Study of the Development of the Regional Assistance Mission to Solomon Islands*, October 2005.

127 Brailey, Malcolm, *The Transformation of Special Operations Forces in Contemporary Conflict: Strategy, Missions, Organisation and Tactics*, November 2005.

128 Stevenson, Robert C, *Not-so Friendly Fire: An Australian Taxonomy for Fratricide*, April 2006.

129 Schmidtchen, David, *Eyes Wide Open: Stability, Change and Network-Enabling Technology*, May 2006.

130 Stevenson, Robert C, *The Human Dimension of the Hardened and Networked Army: The Lessons of Friendly Fire*, June 2006.

131 Blaxland, John, *Revisiting Counterinsurgency: a Manoueverist Response to the War on Terror for the Australian Army*, July 2006. Thai translation published in 2008.

132 Krause, Michael, *Square Pegs for Round Holes: Current Approaches to Future Warfare and the Need to Adapt*, June 2007.

133 Croser, Caroline, *Organising Complexity: Modes of Behaviour in a Networked Battlespace*, November 2007.

Books

Evans, Michael (ed), *Changing the Army: The Roles of Doctrine, Development and Training*, Land Warfare Studies Centre, Duntroon, 2000.

Evans, Michael and Alan Ryan (eds), *The Human Face of Warfare: Killing, Fear and Chaos in Battle*, Allen & Unwin, Sydney, 2000.

Evans, Michael, Russell Parkin and Alan Ryan (eds), *Future Armies, Future Challenges: Land Warfare in the Information Age*, Allen & Unwin, Sydney, 2004.

Parkin, Russell (ed), *Warfighting and Ethics: Selected papers from the 2003 and 2004 Rowell Seminars*, Land Warfare Studies Centre, Duntroon, July 2005.